Befriend a bot
& Spend more time
w family
Enjoy
Jerry [signature]

THE ART OF
AUTOMATION

Discover How AI-Powered Automation Helps
People Reclaim Up to 50% of Their Time at Work

JERRY CUOMO et al

Acclaim for The Art of Automation

"The Art of Automation deftly shows how the future of humans and machines will interact across all industries. Growing adoption will enable a spectrum from human directed to human optional scenarios. This book is a must-read on why automation is an art and not just a science"

—R "Ray" Wang, Founder and CEO of Constellation Research, and 2X Bestselling Author

"Jerry and his guest authors have written the guidebook that clearly illustrates the what, why, and wow of AI-powered automation, with compelling examples from industry experts that have successfully applied it. The Art of Automation leverages Jerry's firsthand insights and provides a practical approach for business and technical leaders to put AI to work to augment your employees to become super-humans and usher in the age of a hybrid workforce."

—Charlotte Dunlap, Principal Analyst, GlobalData Technology

"Jerry and his 'Gang of 7' address the critical question business leaders are attempting to answer – How do I shift my employees focus to higher value work that will ultimately delight my customers? The Art of Automation leverages their firsthand insights and provides a practical approach for business and technical leaders to leverage a proven methodology to improve business performance by making all information-centric jobs more productive"

—Dinesh Nirmal, General Manager, IBM Automation

DEDICATIONS

From: Jerry

To: My dad, who taught me how to be curious, and my mom, who taught me kindness

From: Gang of 7 & Jerry

To: IBM, thank you for "treasuring your wild ducks."

To: Dinesh Nirmal, General Manager, IBM Automation, our sincere thanks for inspiring and sponsoring the Art of Automation project

To: Future survivors, royalties from this book are donated to the American Cancer Society with a matching grant from IBM.

TABLE OF CONTENTS

FOREWORD

AI-powered automation helps people reclaim up to 50% of their time at work to focus on higher value tasks, and that's something we all need.

Foreword by Michael Gilfix

Making AI-Powered Automation Work For You

I'm pleased to have been asked to write the foreword to the book-companion of Jerry Cuomo's podcast, *The Art of Automation*.[1] Automation technology has been around for a long time, but in the past several decades, we've seen an explosion of information work in the enterprise.

When every business is a digital business (which is the case today), the proliferation of data is massive—far greater than humans can cope with alone. And making sense of this mass of data is just the kind of work that classic automation doesn't address. Enter modern artificial intelligence (AI) technology, which can use algorithms to make sense of structured and unstructured data at a speed and scale no human could approach. Making this combination of Automation and AI, which we call "AI-powered Automation," work best for you is, as Jerry says, an art.

Businesses spend billions of hours a year on work that strips people of time and keeps them from focusing on higher value things. AI-powered automation helps people reclaim up to 50% of their time[2], and that's something we all need. If you were to analyze how you spend your day and thought about how much of it is taken up with tasks that add little value. How much of what you're doing would you want on your LinkedIn profile? You wouldn't have your profile highlight mundane tasks like getting all your expense-reports done on time. Instead, you'd highlight the new business you delivered, the IT vulnerability you identified before it impacted thousands of users, and the great value you built for customers. That's one of the ideas behind *The Art of Automation*—you can give yourself and your business hours back thanks to the capabilities now available with AI-powered automation.

THE TECHNOLOGIES BEHIND AI-POWERED AUTOMATION

AI-powered automation includes technologies that together form the building blocks to automate business and IT operations and to collect and interpret data digitally. Among those technologies are concepts that Jerry and others discuss on the podcast and in this book:

- **Natural language processing**, or the ability to make sense of human text and word input,

- **Unstructured data processing**, or the ability to interpret and classify a given document or image,

- **AI for IT Operations (AIOps)**, where AI helps Site Reliability Engineers detect issues early, predict them before they occur, locate the source of the issue, and recommend relevant and timely actions,

- **Robotic Process Automation,** where software robots mimic human interactions with the desktop to perform tasks,

- **Visual recognition technology**, like optical character recognition (OCR), that allows machines to read and understand data in all sorts of formats, and

- Other AI tools that can help act on information, applying machine learning to past outcomes and patterns in order to quickly and accurately support decisions and actions and help predict and avoid potential problems.

These tools and capabilities have become even more critical in the past year, as the pandemic has disrupted industries and ways of working, increasing our reliance on digital as a way to work, learn, communicate, shop, and collaborate. Being responsive to your customers and serving them in new ways is more important than ever, and automation plays a vital part in facilitating that kind of modern customer interaction.

AUTOMATION AND WORKING WITH THE U.S. DEPARTMENT OF VETERAN AFFAIRS

One example we've seen of this is at the U.S. Department of Veteran Affairs (VA)[3]. We worked with them on their claims processing, which, when we started, was an extremely manual, error-prone, and lengthy process. Claims arrived by mail and by electronic communication. The VA required seven hundred employees to sort through this big volume of inbound requests in order to figure out how they should best process those claims. It was extremely error-prone because it was all done manually. Turnaround time was typically seven to twenty days. It not only cost the department a lot of time but the customer experience was also poor.

By leveraging automation, we were able to improve that turnaround time to less than a day, with urgent claims processed in as little as five minutes with little-to-no manual intervention. Automation powered by AI freed up the time of seven hundred VA staff, allowing them to focus on customer relationships and high-value work. And it facilitated a massively improved customer experience.

Modern AI, intelligent automation, AI-powered automation, whatever name you give, it has one essential function: to give people time to focus on the things that matter. Things like relationships, creativity, and decision-making. Things that really make an impact on your customers and the world. Things that help you grow your business and your brand. The kinds of things, in fact, that you would want in your LinkedIn profile. That's the beauty and the art of automation, and this book explores all of its possibilities in detail.

UP NEXT

Make sure you check out *The Art of Automation* podcast, especially Episode 10, in which Jerry and Mike discuss the "business of automation" and how AI-powered automation technology gives businesses back "the gift of time," allowing people to spend their days on the things that matter.

Now, fasten your seatbelts; the book is about to take off.

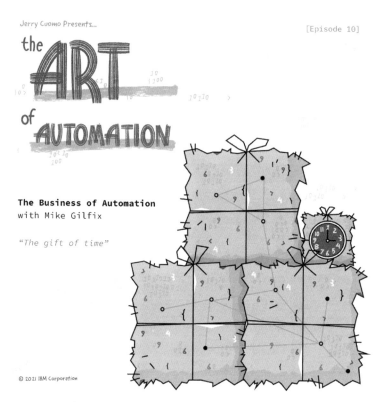

Foreword: Figure 1 – The Art of Automation podcast Episode 10

PREFACE

Jerry Cuomo and a distinguished cast of co-authors present *The Art of Automation*, providing leaders with an understanding of how AI-powered automation can hyper-accelerate digital transformation.

Preface by Jerry Cuomo

WHOM IS THIS BOOK FOR?

For enterprises large and small, the pandemic has meant an acceleration in digital transformation by months or even years. There is no going back. Digital transformation is the process of using digital technologies to create new, or modify existing, business processes, culture, and customer experiences to meet changing business and market requirements[4]. The move to digitization has most certainly accelerated, and the benefits will be permanent. Automation technology stands out as a top enabler for digital transformation, sought after by business and technology leaders. The primary intent of *The Art of Automation* is to provide these leaders with a fundamental understanding of how AI-powered automation can hyper-accelerate this transformation and help their business grow.

This book is a practical guide for leaders who need to understand and evaluate how automation technology can transform their organization, business processes, and models. It combines both an aspirational and pragmatic overview of automation technology, its core capabilities, and the value they generate from a business perspective. It describes various real-world examples, implementations and approaches, with industry-specific and cross-industry use cases.

Leaders who are unsure of where to start their automation journey can use the examples in this book as a guide for selecting projects and technologies with techniques designed to speed up the automation of "bite-size" activities that deliver immediate return on investment (ROI) without having to wait until the entire end-to-end process is automated.

FROM PODCAST TO BOOK

In this section, I cover the inspiration behind this book. It all starts with customers. Over the past year, we have been hearing from IBM customers at an increased rate, pace, and volume that artificial intelligence (AI) and automation are two of their highest priorities of investment coming into 2022. We in the IBM Software unit have been listening and feel strongly that we have much to offer in the form of advanced products, services, and consulting. But, given the heightened interest, we felt we can do even more.

Specifically, IBM has made AI-powered automation a top priority in 2022 and beyond. To that end, we have formed a business unit—IBM Automation—focused on further growing capabilities in the area of automation to serve these customers' needs.

In this pursuit, I started talking with many of IBM's subject matter experts across Watson Research, Consulting, Data & AI, Business Automation, IT Automation, Integration, and everything in between. I was floored by their knowledge, energy, and passion for the AI-powered automation topic and how their breakthrough thinking could be used to make a difference for our users. I could not keep these experiences to myself and had to share their energy, up close and personal, with our customers and the industry. This is where the idea for *The Art of Automation* podcast came in. I called

Ethan Glasman, a colleague at IBM, who I knew could help. His advice was to keep the topics focused, short (fifteen minutes or less), and inviting to ensure the personalities of the guest shone through.

My initial set of experts did not disappoint and included the likes of Rama Akkiraju, IBM Fellow and *Forbes* Top 20 Women in AI Research on the AIOps topic, and Rania Kalaf and Ruchir Puri, a pair of distinguished IBM researchers, on the topics of Digital Workers and AI for Code.

Once word got out, I started hearing from listeners who also wanted to hear from some industry-aligned experts. To date, I've been lucky to have industry luminaries like Claus Jensen, CIO Memorial Sloan Kettering on Automation in Healthcare, and Carol Poulsen, CIO of Cooperators on Automation in Insurance. As of the writing of this preface, we've completed Season 1 with twenty-three episodes under our belt and over eleven thousand downloads and counting!

And now, here it is… *The Art of Automation* book. We decided to create this book to capture this important moment in the IT industry, where automation and AI join forces to free humans from their mundane tasks such that we can all focus on higher value, post-pandemic tasks that matter more to one's business.

Publishing in this medium (which includes an e-book) allows the content to be searchable on the internet, while also allowing my co-authors to go into more detail, including references on how to find more information to progress one's automation journey.

Each chapter celebrates the unique voices of my colleagues in a subject that they are experts in.

WHAT IS COVERED IN THIS BOOK?

Our goal is to cover all the key topics with which you need to be comfortable in order to positively impact your organization as it evaluates and implements AI-powered automation technology. Even if you are already familiar with the basics, the early chapters will reinforce your understanding of the most important concepts (e.g., closed-loop automations) and technologies (e.g., Robotic Process Automation [RPA] and AIOps) and explore general use cases. As you dive deeper, you will be systematically introduced to specific topics with details that will enable your organization to successfully implement solutions that leverage AI-powered automation. The book also provides industry-specific points of view on automation through a series of chapters that capture interviews with luminaries in the fields of healthcare, insurance, retail, and financial services.

Feel free, however, to jump directly to the chapter that most directly impacts your current role and answers your most immediate questions. You will also find references for further study throughout the chapters to fill in any gaps or provide more detail, depending on your level of experience or organizational role.

Several chapters are co-written by some of the industry's most notable experts in AI and automation. Many of these experts can be heard on a corresponding episode of *The Art of Automation* podcast. See the "Up Next" sections that conclude each chapter.

The following illustration provides a high-level map of the chapters, which is an easy way to understand the structure of this book.

The Art of Automation

Into	This chapter
Business	Digital Workers | RPA | Intelligent Documents | Process Mining
IT	Observability | AIOps | APIs
Use-cases	Healthcare | Insurance | Banking | Retail | Sea | Weather
Future	Yesterday, Today, Tomorrow
Bonus	The Art Behind the Art of Automation

Preface - Figure 1 – The Art of Automation: Chapter Map

The book starts with a general introduction to AI-powered automation, followed by a series of chapters, written by subject matter experts, that deep-dive in the areas of business automation (chapters 2–5) and IT automation (chapters 6–8).

Following these are chapters (9–14) that cover industry use cases. These chapters take the form of enhanced transcripts from the actual episodes of *The Art of Automation* podcast and feature the perspectives of podcast guests who share their experience with AI-powered automation within the context of their industry.

Chapter 15 is a summary that looks at automation past, present, and future. And last but certainly not the least, Chapter 16 concludes with a fun look at the unique cover art of each podcast episode and the stories from the artists that created them. Here is a short summary of each chapter.

Chapter 1: Introduction

Many companies are declaring this year as "The Year of Automation," seeing automation as an essential part of how we address our new normal.

This chapter introduces *The Art of Automation* book and the "art of the possible" of AI-powered automation in the enterprise. Jerry elaborates on the concepts of automation and self-driving cars, and how these relate to the future of enterprise technology.

By: Jerry Cuomo

Chapter 2: Robotic Process Automation (RPA)

RPA is made up of three core technologies—workflow automation, screen scraping, and AI. It is the unique combination of these three technologies that allows RPA to solve the productivity challenge of manual desktop tasks. Allen describes how RPA automates repetitive human tasks by mimicking human actions and how this can save time in tasks such as reading documents and interacting with applications.

By: Allen Chan and Jerry Cuomo

Chapter 3: Process Mining

Process mining gives you the tools and methodologies that you need to unlock the data that shows how processes work, how people do their jobs, and where problems are coming from. Harley explains process mining takes a data-driven approach to translating an enterprise's desire for improved operations into actions and automations that can actually be implemented.

By: Harley Davis

Chapter 4: Digital Employees

A digital employee is a new type of automation software that makes it easy for any human worker to automate their day-to-day tasks without being a programmer. Salman describes how digital employees are the next evolution of Conversational AI. Predicted by Gartner and realized in IBM's Watson Orchestrate, a digital employee extends the conventional concepts of RPA, ChatOps, and Digital Workers providing a multimodal experience and possesses "skills" with deep context and with a wide range of enterprise app integrations to tackle more complex virtual assistant use cases.

By: Salman Sheikh

Chapter 5: Intelligent Document Processing

Organizations are spending more and more time manually processing documents, where we can't just blame the poor image quality of the fax machine. Eileen discusses why document processing is so manually intensive and what makes it so difficult to automate. Eileen describes how AI/ Deep Learning can mimic the human brain to understand the unstructured information found in an enterprise's wide variety of documents. She shares an example from the insurance industry that shows the enormous business value of Intelligent Document Automation.

By: Eileen Lowry and Jerry Cuomo

Chapter 6: Observability

Observability is the "data source" of automation, maximizing the amount of visibility into your enterprise operations with the least amount of effort. Jerry discusses how any major IT role benefits from the real-time insight gained from observability software and that observability solutions are evolving to benefit both business and IT leaders in a way that aligns technology efforts and investments with strategic business objectives that simply deliver better results faster—and this is what observability is all about.

By: Jerry Cuomo

Chapter 7: AIOps

We envision fully instrumented, observable, self-aware, automated, and autonomic IT operations environments in the future. AI can help us get there. Rama describes the incredible automation potential that AI is bringing to IT operations management and introduces the idea of a "digital teammate" and explains how it all comes together in "ChatOps."

By: Rama Akkiraju

Chapter 8: APIs

Rob explains the relationship between APIs, integration, and automation and why "you can't automate an enterprise, unless that enterprise is programmable through APIs." Rob describes how, in order for an enterprise to go faster, it needs quick and easy access to all relevant data and events, and how the best way to do this is to make APIs ubiquitous across the company.

By: Rob Nicholson

Chapter 9: Healthcare

In a conversation on automation in the healthcare industry, Jerry Cuomo and Claus Jensen from the Memorial Sloan Kettering Cancer Center discuss the important role automation is playing in the healthcare industry and why the combination of AI and human doctors outperforms either in isolation. Claus shares good (and bad) examples of how AI can be used in healthcare, as well as what innovations he sees on the horizon in the coming years.

Interview by: Jerry Cuomo

Chapter 10: Insurance

Jerry Cuomo and Carol Poulsen from The Co-operators Group on Automation in the Insurance Industry discuss the important role automation is playing in the insurance industry and how companies are moving from standard risk mitigation to AI-powered risk prevention. Carol describes how automation is allowing insurance to put customers at the center, instead of at the end of a process. She also shares what innovations AI could bring to insurance in the coming years and how those innovations could shift risk and save you money.

Interview by: Jerry Cuomo

Chapter 11: Retail

Talking about automation in the retail industry, Jerry Cuomo and Tim Vanderham from NCR Corporation discuss how progressive retailers are beginning to use automation to transform the customer's shopping experience—a transformation that has been catalyzed by the ongoing pandemic. Tim shares automation examples from self-checkout to online ordering and explains how this goes beyond classic automation to edge computing, computer vision, internet of things, and even cryptocurrency. They close by jumping the tracks to AI and diving into how, out of necessity, retail companies are utilizing AI to filter through millions of events to isolate outages and create a more personalized experience for their consumers.

Interview by: Jerry Cuomo

Chapter 12: Financial Services

This chapter entails a conversation between Jerry Cuomo and Oscar Roque from Interac Corporation on automation in the financial services industry. They discuss how AI-powered automation in the financial sector can positively impact multiple industries and become a crossroads for automation to create new value networks. Oscar discusses the challenges of "ingesting a world of data" both across payments networks and all other industry networks that are touched by payments. He also shares that the outcome of automation can lead to the gift of time allowing skilled works to focus on solving higher order (community, social, and business) issues.

Interview by: Jerry Cuomo

Chapter 13: Automation and the Weather

In a conversation, Jerry Cuomo and Lisa Seacat DeLuca—distinguished engineer, author, and one of the most prolific inventors in the history of IBM (with over eight hundred patents) on automation and the weather—discuss how AI-powered automation helps businesses understand the environment and how they can use that data to tackle challenges related to climate change, sustainability, and everyday business operations. Lisa shares an example from the airline industry, where flights still use a paper-based flight manifest that a pilot must sign, and she explains how much faster and more accurate the process could be if reinvented with AI-powered automation.

Interview by: Jerry Cuomo

Chapter 14: Automation at Sea

Jerry Cuomo and Don Scott, director of engineering at Submergence Group, and the mastermind behind the Mayflower Autonomous Ship, talk about automation at sea. They discuss the unique application of automation and AI at sea, specifically around the Mayflower Autonomous Ship project, which is attempting to become the first entirely automated vessel to traverse the Atlantic Ocean. Don explains how this "AI Captain" operates and makes decisions while at sea. He also elaborates on safety, trust, and international marine regulations and how AI explainability is such an important part of the Mayflower Autonomous Ship (or any other AI system).

Interview by: Jerry Cuomo

Chapter 15: Yesterday, Today, and Tomorrow

Jerry Cuomo and Ed Lynch, vice president of IBM Digital Business Automation, converse on the automation of yesterday, today, and tomorrow. Together they take a step back and discuss the concept of Enterprise Automation as a whole, specifically, where we are today and how we got here. Ed explains that automation is everywhere, not limited to one specific industry. Ed then elaborates on the future of enterprise automation, and how it inevitably will be centered around AI and augmenting human work. They close by determining that automation all comes down to the human beings and that the best place to begin your automation journey is with the data.

Interview by: Jerry Cuomo

Chapter 16: The Art behind the Art of Automation

Like any good television show, musical, or other piece of media, the success and popularity of *The Art of Automation* podcast has quite a bit to do with the support staff who make the show happen from a distance. The cover art for *The Art of Automation* podcast is recognizable anywhere. *The Art of Automation* has worked with some of IBM's most talented designers and visual artists to create the cover art that is now recognized by listeners across the world. In this unique chapter, Ethan uses it to "pay attention to the people behind the curtain," with interviews with the designers that produced the cover art for the podcast.

By: Ethan Glasman

Chapter 17: Ciao

Goodbye for now? This chapter wraps up any remaining loose ends. Setting a context for potential topics covered in this book, this chapter reviews topics that the author wishes he could have covered, several of which are topics covered in past episodes of the podcast. Most importantly, a LinkedIn account is introduced, Art of Automation – DJ, as a means for readers and listeners to stay connected, allowing for comments, surveys, and additional material related to AI-powered automation in the enterprise. And last, but certainly not the least, the book concludes with a series of acknowledgments. Well… that's all folks.

By: Jerry Cuomo

MEET THE AUTHORS

Author

Gennaro (Jerry) Cuomo is a prolific contributor to IBM's software business, producing products and technologies that have profoundly impacted how the industry conducts commerce over the World Wide Web. He is most recognized as one of the founding fathers of WebSphere Software, whose innovations defined WebSphere as the industry-leading application server serving over eighty thousand customers.

Cuomo has filed for over a hundred US patents and has been cited over three thousand times. Jerry's most visible patent is the first use of the "Someone is typing…" indicator found in instant messaging applications.

At IBM, Cuomo has led projects in the areas of AI-Powered Automation, Blockchain, APIs, cloud computing, mobile computing, Internet of Things, web server performance and availability, web security, web caching, edge computing, service-oriented architecture, and REST. Cuomo is the co-author of the book *Blockchain for Business*, which illustrates how blockchain technology is re-imagining many of the world's most fundamental business interactions and opening the door to new styles of digital interactions that have yet to be imagined.

Cuomo is currently the VP and CTO of the new IBM Automation business unit, where he is driving the technical strategy for AI-powered Automation, and he hosts *The Art of Automation* podcast.

Guest Authors

You may have already seen a few references to the "Gang of 7" as a loving way to refer to the seven guest chapter authors of this book. This term is a play on the infamous "Gang of 4," who were the IBMers that wrote a highly influential technology book called *Design Patterns*[5]. Here is a short bio on the automation subject matter experts whom you will have the pleasure of hearing from in this book.

Rama Akkiraju is an IBM fellow and CTO for AI Operations portfolio of products at IBM. Rama was named among the "Top 20 Women in AI Research" by *Forbes* magazine in May 2017, among the "Top 6 A-team for AI" by *Fortune* magazine in July 2018, and among the "Top 10 Pioneering Women in AI and Machine Learning" by Enterprise Management 360 in April 2019.

Rama served as the president for the International Society for Service Innovation Professionals (ISSIP) in 2018. She holds a master's degree in computer science and an MBA from New York University's Stern School of Business. Rama has co-authored six book chapters and over a hundred technical papers. She has over thirty-five issued patents and more than twenty others pending.

Allen Chan is an IBM distinguished engineer and CTO for Digital Business Automation. He is responsible for the IBM Cloud Pak® for business automation, including capabilities such as business process management and case management, decisions, content management, application designer, and robotic process automation.

Prior to that role, Allen had held various technical leadership roles in IBM BPM, such as the chief architect for Workflow (IBM BPM and Case Manager) and Blueworks Live. He is the holder of multiple patents in Canada, United States, and China, and he is the author of multiple articles around business processes, automation, and applied AI in automation.

Harley Davis is a software industry executive with international start-up and senior-level management experience in software development, consulting, and sales. He is currently leading automation intelligence and the IBM France R&D Center at IBM, helping companies drive business agility in their operations with the world's leading platform helping organizations apply AI and other technologies to improve processes, decisions, and content management.

Ethan Glasman is a technical content creator based in San Jose, California. He is currently the producer of *The Art of Automation* podcast as well as the technical communications lead for IBM Automation. His background is in applied mathematics and theater, and he enjoys combining technology and creativity to make engaging stories that are accessible to anyone.

Eileen Lowry is the vice president of product management for business automation—content services portfolio in IBM Automation. Prior to leading product management for content services, she led the IBM Blockchain team that designed, built, and deployed blockchain applications on top of the IBM Blockchain Platform and Hyperledger technologies.

Eileen has a strong business acumen and an understanding of enterprise software, with experiences in product management, technical services, mergers and acquisitions strategy, business and corporate development, financial planning and analysis, pricing strategy, and software development.

Rob Nicholson is a distinguished engineer with more than twenty-five years' experience leading innovation in software development teams, with a passion for technology and the development teams who create it. He has led IBM's Integration technology strategy as the CTO for APIs and Integration, responsible for world-renowned integration software, including MQ and Integration Bus, as well as new modern integration technology including Event Streams and API Connect.

Rob is an IBM "Master Inventor" with more than fifty issued patents broadly across microelectronics, software, firmware and business processes, and AI-infused integration technology.

Salman Sheikh is the program director of product management at IBM Automation. In his eight years at IBM, he has served in various product management and design leadership roles defining product visions and execution strategies to deliver award-winning, market-leading enterprise solutions.

Most recently, Salman is responsible for bringing Watson Orchestrate to market as a breakthrough AI technology that puts automation into the hands of employees to automate day-to-day tasks.

UP NEXT

Each chapter in this book concludes with an "Up Next" section. The purpose of this section is to keep the storyline of the book together by tying the past and current chapters with the chapters to come. This section also provides a pointer to an *Art of Automation* podcast episode that corresponds with the subject matter covered in the chapter. I also couldn't resist including an image of the cool cover art that was created for that episode. So, following that recipe for the preface, make sure you check out the "Bonus" episode of *The Art of Automation* podcast in which I introduce this book. Listen in and you will hear the excitement in my voice as I describe a summary of what you are about to read.

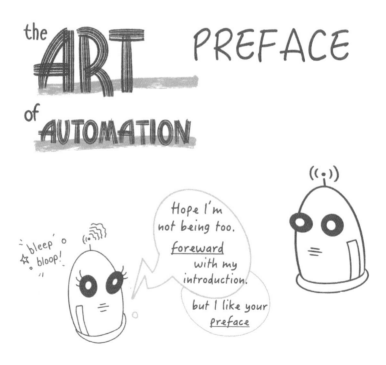

Preface - Figure 2 – The Art of Automation *Podcast "Book" bonus episode*

Chapter 1

INTRODUCTION

Many companies are declaring this year as "The Year of Automation," seeing automation as an essential part of how we address our new normal.

Chapter Author: Jerry Cuomo

COVERED IN THIS CHAPTER

- Automation definition

- Process: Discover, decide, act, and optimize

- Examples of business and IT automations

- Debunking automation myths

- IBM Automation overview

HERE WE GO...

This introductory chapter to the Art of Automation provides the fuel that powers the chapters to come. Starting by establishing AI-powered automation as a key aspect to the new normal emerging from the world-wide pandemic, you will learn about the process of automation and how it is structured around a feedback loop involving phases of discover, decide, act, and optimize. Applying artificial intelligence (AI) during each of these phases propels automation to go beyond what was doable before, including understanding opportunities to predict versus react and see actionable patterns in noisy unstructured data.

This chapter outlines the styles of tasks that can be automated across an enterprise while also clearing the air by debunking popular myths about AI and automation that readers need not be burdened with while reading the book.

"THE YEAR OF AUTOMATION"

Many companies are declaring this year as "The Year of Automation," seeing automation as an essential part of how we address our new normal.[6] The COVID-19 pandemic has changed the dynamics of business and how we work.

For example, YouTube recently said in a blog post that with fewer people in its offices around the world, automation software is doing more of their video content moderation. "We have started relying more on technology to help with some of the work normally done by [content] reviewers," the company said. "This means automated systems will start [curating] content without human review."

The ideas around automation are simple, are "tried and true," and have been applied to business since the very beginning of the industrial era. So, why do we feel the time is right now for automation?

Businesses today are all-in on being digital. The pandemic has accelerated digital transformation, with digital means to do just about everything from ordering a pizza to telemedicine to our Zoom-enabled workplaces—all *digital-powered*. As every aspect of a business becomes digital, the doors open for automation. In a sense, businesses, like cars, have become computers that can be programmed, and automation is the software that can propel a business to have "autopilot" or "auto-assist" modes. Digital-powered then sets the table for *AI-powered*.

Today, we have technology that can greatly change the "state of the art" of automation, with artificial intelligence (AI), machine learning, computer vision, and natural language processing. AI has given birth to new ways to automate things that were difficult to automate before. AI, in the form of machine learning models, applied to IT automation with AIOps can predict the risk associated with making a change to your application, avoiding costly outages. Similarly, on the business side, natural language understanding with robotic process automation can put automation in the hands of every business user to automate time-consuming and error-prone data entry tasks.

So, the Year of Automation is ushered in by a pandemic-accelerated, digital-powered revolution and is being paired with AI-powered technology, setting a big stage and placing intelligent automation at its very center. As such, many industry analysts have declared automation—or extreme automation, hyper-automation, etc.—as the most critical technology trend to act on now.[7]

ART OF AUTOMATION FUNDAMENTALS

Automation as an "art form" for business could be simply viewed as a two-step dance. As discussed in the previous section, our new normal is driving client demand to automate processes that *eliminate repetitive* and monotonous tasks. This is the first step in the dance and enables the second step, which is to *augment humans* to produce super-human results more rapidly.

Here is a closer look at these two simple yet fundamental aspects of the Art of Automation:

Task elimination targets simple, repetitive tasks across business and IT. Automating these tasks will free employees up to do more thoughtful work. For example, with closed offices keeping many of its workers away, PayPal has turned to chatbots, using them for a record 65% of message-based customer inquiries in recent weeks. "The resources we are able to deploy through AI are allowing us to be more flexible with our staff and prioritize their safety and well-being," PayPal said in a statement.[8]

Task augmentation supports, speeds up, and increases employee efficiency. For example, with the increased use of online services during the coronavirus pandemic, AI-powered customer service agents can allow a single agent to help more users, decrease service queues, and increase customer advocacy. AI is used to gauge user intent and capture information and the nature of the problem the customer is asking the company to solve. An automation workflow can then examine possible resolutions with, or possibly without, engaging a human. That being said, the most powerful form of task augmentation is when humans and AI systems work hand in hand in achieving the desired outcome.

It is tempting to some businesses to stop after the first step and declare victory because of cost savings associated with task elimination. However, the true Art of Automation crescendos with augmenting humans with automation software. On Episode 3 of *The Art of Automation* podcast, Claus Jensen, chief digital officer at Memorial Sloan Kettering Cancer Center, commented that automation software won't replace doctors. Instead, it's really the combination of artificial intelligence and human doctors which, he stated, "outperforms either in isolation." Claus further exemplified his comment with how doctors are using AI and computer imaging to enhance X-ray photos, highlighting anomalous structures, which then allows a doctor to discern and make a diagnosis, which would not have happened if the doctor used his eyes alone. A great example of the Art of Automation in action!

AUTONOMOUS VEHICLES AND THE PROCESS OF AUTOMATION

Automation has graduated to hyper-automation. We see it on the streets, with *self-driving autonomous* vehicles. This is truly a marvel of modern ingenuity. Leveraging the best of artificial intelligence, internet of things, and cloud computing, the automotive industry has set the "bar high" with this accomplishment.

Test driving a Tesla Model 3 with its autopilot features is a fun way to witness the process of automation at work while gaining an appreciation for how Tesla has taken automation to an extreme with advanced technology. To keep with the autopilot theme, we will use the following figure as a guide to examine the process of automation with an eye towards how we might apply this same process to create an autonomous business.

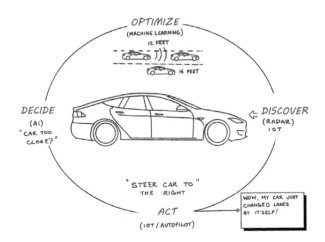

Chapter 1 - Figure 1. Autonomous vehicle automation process

Think about the "wow" moment when the car changes lanes by itself. This outcome is achieved through a process described as a set of tasks. This process of automation involves *discovering* data from sensors, including radar. As data is discovered, it is analyzed and correlated, such that another car might be recognized ahead. Then *decisions* are made. Is the car ahead too close? Based on a tolerance threshold, a proactive decision to trigger an *action* to slow down or switch lanes occurs. The discovery, decisions, and actions occur simultaneously and are continually *optimized* against past and present data to predict future actions, forming an AI-fueled closed-loop automation system.

While there are dozens of variations of closed-loop automation (remember MAPE loops?[9]), we prefer to use this simple 3-plus-1 methodology described in the above example: *Discover, Decide, Act,* and *Optimize.*

Discover involves collecting, organizing, and classifying structured and unstructured data that flow through your enterprise. *Structured data* comprises clearly defined data types whose pattern makes them easily searchable. *Unstructured data*—"everything else"—comprises data that is usually not as easily searchable, including formats like audio, video, and social media postings.[10]

AI is used to understand relationships and correlation, derive deep insights, find gaps, and establish baseline Key Performance Indicators (KPIs). Patterns recognized in data can be classified as bottlenecks, hot spots, anomalies, or outliers. They provide a context for *decisions* to optimize business and IT processes, using historical performance and predictive analytics to help you deal with variations. Automation occurs when decision-logic triggers *actions* that can proactively alert, tune configuration, make API calls, and run programs that once required employee intervention. The "plus one" is *optimize.* Optimization is +1 because it's occurring throughout all three phases, providing a feedback loop that continually discovers new data patterns, improves decisions based on past actions, and provides explanations and evidence that allow actions to be performed autonomously with confidence.

The autonomous vehicle example is inspirational. Now, how can we apply technology to similarly transform your enterprise to become an *autonomous enterprise*, with process automation and advanced technology to establish insights and automate actions?

AI-POWERED AUTOMATION IS ENTERPRISE AUTOMATION 2.0

To achieve these results, we are actively working to advance automation technology towards AI-powered automation, which we declare to be Automation 2.0 or, as coined by Gartner, hyper-automation. AI-powered automation is defined as a continuous closed-loop automation process where data patterns are *discovered* and analyzed, such that *decisions* on insights from the data can be translated into automated *actions*, with AI providing proactive *optimizations* during each stage of the process. AI-powered automation uses actionable intelligence to deliver IT and business operations with speed, lower cost, and improved user experience. The next section examines these four stages, illustrating how AI is transforming at each of these stages.

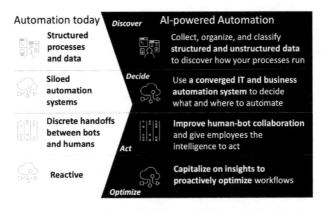

Chapter 1 – Figure 2: Evolution to AI-Powered Automation

Discover

Better understand and classify unstructured data and processes so you can lessen the burden of manually analyzing and orchestrating actions.

Without AI, data discovery associated with automation is mostly limited to structured processes and structured data. While AI is often used to understand structured data, unstructured data is trickier because it is inherently noisy and often slows down the automation process. With the use of machine learning (ML), models are produced to cut through, tease out, and detect patterns in the noisy data. For example, with a properly trained classifier model, unstructured documents can be "structurally classified" as an invoice or insurance claim. Similarly, alerts from an IT system can be grouped and matched to a specific trouble ticket. With AI, the *discovery process is no longer blocked by lack of structure*; it uses AI intelligently to move from discovery to decision-making.[11]

Decide

Combine the precision of IT automation with well-defined methodology of business automation so you can automate faster and with more accuracy in both IT and business.

AI-powered automation aims to comprehensively provide *a converged business and IT automation system* that operates *across a broad range of labor types*, including business workers, solution architects, software engineers, IT operations, site reliability engineers (SRE), security and compliance engineers. By discovering data patterns across business and IT, *decision-making* can now be more impactful versus systems that are siloed to specific parts of an enterprise. An example of this is deciding whether or not to release a new piece of source code with insight gained by correlating activity across software development and IT operations. In this case, changes to source code and configuration during development can be matched against incidents happening in a running IT system to predict risk associated with future changes to that code or configuration. By applying AI to automation, we are greatly improving the speed with which an enterprise can react to new patterns discovered and make decisions to prevent incidents versus reacting after they've taken place.

Act

Engage software bots more naturally and collaboratively so engagements become more self-service and productive.

The automation process is further differentiated in how automated actions are carried out. The gold standard in automating actions is robotic process automation (RPA) technology. With power from AI, we are evolving RPA from simple robotic scripts to becoming a tech that is more like a *digital twin in the workplace.* A digital twin is a virtual model of a process, product, or service. This pairing of the virtual and physical worlds allows actions to be simulated in order to head off problems before they even occur, prevent downtime, and develop new opportunities. Furthermore, Automation 2.0 uses advanced natural language processing to produce a more collaborative relationship between AI and employees to produce a hybrid workforce where automating actions in an enterprise becomes a "team sport."

Optimize

Predict potential incidents earlier so systems can proactively resolve issues before they impact normal operations.

Optimizations are continuously applied during discover, decision, and action phases, capitalizing on new insights to autonomously enhance business and IT operations through closed-loop feedback. In Automation 2.0, *optimizations move beyond reactive to predictive and proactive.* With an end-to-end view of data across business and IT, AI-powered automation can anticipate fluctuations and help avoid overreacting. For example, by combining structured and unstructured properties of historical change and incident records from enterprise IT, linkages can be extracted between change incidents to create empirical evidence as new inputs to a change risk model. As new changes are being rolled out by IT, real-time proactive alerts can be issued based on predictions that illustrate why these changes are high-risk based on past evidence. Gartner Market Guide for AIOps platforms[12] declares this proactive style of risk management as the most sophisticated stage of automation.

Optimization is fueled by Machine learning and artificial intelligence (AI) enabling new forms of intelligent automation. As the software "learns," the more adaptable it becomes. These technologies open the door for automation of higher order tasks as well, rather than just basic, repetitive tasks. Automation is not just about automating those tasks humans are doing today but also about realizing new potential opportunities.

As data sets become more thorough and available, and as software draws on more sources and synthesizes more data points, contextual information in human decision-making will only improve. Machine learning, then, will serve as a supplement to human knowledge.

EXAMPLES OF BUSINESS AND IT AUTOMATIONS

Hundreds of thousands of discrete tasks make up the thousands of activities that drive the hundreds of processes within a digital enterprise; each individual task is an automation opportunity.

So, where does one begin? Developing an automation strategy in advance enables organizations to optimize investments by striking a balance between the difficulty of automating a task with its potential increase in efficiency. An IBM Institute for Business Value study[13] showed that "one out of two executives using intelligent automation have identified the key processes within their organization that can be augmented or automated using AI capabilities."

Analyzing work activities is the most accurate way to assess the potential for automation. The American Productivity and Quality Center (APQC)[14] publishes a list of almost 1,100 cross-industry activities that compose three hundred core enterprise processes. These processes hold the greatest potential to gain greatest efficiency.

One might correctly conclude from these studies that just about "anything and everything" is in-play to be automated. That said, the following section takes a closer look at the sorts of tasks that can be automated in an enterprise across three specific classes of automations, which include *IT operations* (ITOps), *software delivery*, and *business* categories, as follows:

IT automation examples

A closer examination of 884 ITOps automation use cases from IBM Consulting illustrated four primary categories in labor can be offset by automation:

- *Service request management*: Access provisioning, org changes, compliance requirements

- *Event management*: Monitoring, alerting, remediation of commonly occurring events, self-service

- *Access management*: Provisioning, revoking access, bulk access, onboarding/offboarding

- *Service desk and ticketing*: Similar ticket identification, report generation

Software delivery automation examples

On the software delivery side, we see automation opportunities where gaps fall into these additional categories:

- *Application performance management*: Hot spot analysis, impact analysis on environment (processor, CPU, memory)

- *Compliance management*: Conformance to industry regulations (including NIST PCI, GDPR, HIPPA)

- *Application code vulnerability management*: Open-source code provenance

- *Software quality management*: Anti-pattern detection, test coverage, release risk assessment

Business automation examples

From our business automation experience, we know automation gaps fall into four categories:

- *Workforce management*: Email marketing, talent acquisition and employee recruitment, customer service chatbots

- *Case management*: Routing case ownership with queues, assigning cases automatically, responding to customers automatically, escalating cases when necessary

- *Policy and compliance*: Reporting compliance status and audit information, continuous verification of compliance requirements, managing risks and catching potential weaknesses

- *Process handling*: Invoice processing, reconciliation, exceptions, and approval

In the chapters to come in this book, these twelve categories of automation will be further explored both from a technology and an industry usage perspective.

DEBUNKING AUTOMATION MYTHS

While the world is generally enthusiastic about the positive impact AI-powered automation will have on an enterprise, I also hear other sides of the argument. While "questioning" is a natural part of the scientific method, I do see a handful of questions asked repeatedly. In this section I respond to three such questions that I've cataloged as "automation myths." While there is always an element of truth to such myths, I hope you we agree that the balanced view provided in the sections to come draws a conclusion that there is little to fear in embracing AI-powered automation now. There is also a deeper dive into the myths related to "the impact automation will have on employees" in the "Digital Employees" chapter of this book.

Speed of automation

Myth: The automation process is a slow one.

Truth: The automation process of discover, decide, act, and optimize might lead one to think that automation is a sequential and time-consuming process. While it is absolutely true that perfecting the automation process can take weeks or months, there are fast paths forward. For example, the use of RPA and low-code development are all designed to speed up the automation of "bite-size" activities or processes so customers can get immediate ROI without having to wait until the entire end-to-end process is automated. The fast turnaround time also allows business and IT to fail fast by iterating quickly and responding in real time to external forces.

Automation for everybody

Myth: Automation is only for the likes of data scientists.

Truth: AI-powered automation does not require everyone involved to be a data scientist and understand the intricacies of machine learning, training, etc. On the contrary, the use of AI enables automation technology to reach the general business user population, in addition to the IT developer, highly skilled knowledge workers, and, of course, the data scientist. Users across the enterprise benefit from pre-trained models that were prepared by experts—allowing for immediate use—without requiring deep AI skills. Delivering AI-powered automations using natural language and chatbots creates an environment where the automation system meets users where they work, the way they work. This provides a more natural interaction, enabling more workers across an enterprise to both contribute and benefit from automation.

Automation and jobs

Myth: Automation will replace humans and take jobs away.

Truth: One of the greatest myths we hear about automation involves concerns that somehow software will replace humans and take away jobs. On the contrary, automation will add jobs. There are two key aspects to AI-powered automation. Automation is about freeing up labor, which then allows these workers to focus on what matters most to business.

It's about doing more meaningful work—work that will result in a better customer experience. It's a shift of focus towards more productive work, not the removal of work. In fact, much of the evolution of technology goes in this path. The technology industry has always created many more jobs than it has eliminated, says 140 years of data. A study of census results in England and Wales since 1871 finds "the rise of machines" has been a job creator rather than making working humans obsolete.[15]

IBM AND THE FUTURE OF AI-POWERED AUTOMATION

The chapters to follow in this book are written by IBM subject matter experts in AI-powered automation. As you will see, they don't just "talk the talk," they also "walk the walk," meaning that much of their skill and experience in automation has been gained by working on automation products, technology, and customer engagements.

Hence, as you read this book, you will sometimes hear references to products and technology that have influenced their points of view on AI-powered automation. Therefore, this section provides a thirty-thousand-foot view of the IBM Automation offerings to assist you in seeing how all their individual slices fit into the bigger pie. This section is not meant to be an IBM advertisement, but instead is here to provide a backdrop into how and why the technology chapters of this book are organized the way they are.

IBM's approach to AI-powered automation takes the form of a converged business and IT automation system with an ability to continually optimize by discovering, deciding, and taking action as a means to automate processes across an enterprise. With this end-to-end view of automation, we are taking a bold step towards creating a hybrid workforce where your employees—collaborating with their digital twins—can gain deeper efficiencies across their business and free up time and money to focus on new business opportunities.

The IBM Automation offerings have three suites of capabilities:

1. Business Automation

2. IT Automation

3. Integration

The three offerings suites under IBM Automation are delivered as IBM Cloud Paks®, which are containerized software that uses Red Hat OpenShift as the means by which the software can be run and managed on any cloud—public or private.

Each Cloud Pak has the best of breed automation unto its own. However, when the Cloud Paks are deployed together, their value is multiplied. For example, when Business Automation and AIOps are deployed together, business and IT events can be linked such that a business user can, in real time, assess the business impact (e.g., time, cost, customers affected) of an IT incident. They can then use automation (RPA bots) to notify impacted users of the situation or to take corrective action to reduce the impact before users even realize an anomaly has occurred.

The following figure illustrates the IBM Automation offering structure. The diagram is tagged with a set of "location pins" that signify the chapter number in this *Art of Automation* book that covers that piece of the IBM Automation offering.

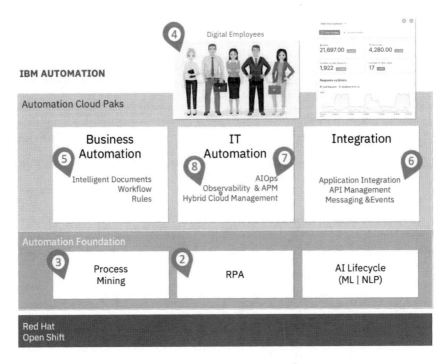

Chapter 1 – Figure 3. IBM Automation offering structure

Business Automation: Providing technology solutions in the domain of business automation, targeting line of business users as the predominant user persona. These capabilities are included in the IBM Cloud Pak® for Business Automation, including the following:

- Process automation and orchestration (workflow)

- Decision processing with business rules

- Intelligent document processing

IT Automation: Providing technology solutions in the domain of IT Operations Management, targeting ITOps, DevOps, and SREs as the primary user personas. These capabilities are included in the IBM Cloud Pak® for Watson AIOps, including the following:

- AI Operations (AIOps)

- Multicloud management (MCM)

- Observability and application performance management (APM)

- Application resource management (ARM)

Integration: Providing technology solutions in the domain of system integration, targeting integration engineers, ITOps, and developers as the primary user personas. These capabilities are included in the IBM Cloud Pak® for Integration, including the following:

- Application integration (App Connect)

- API technology (API Connect)

- Messaging and events (MQ and Event Streams)

Automation foundation: Each of these Cloud Paks uses a common set of automation services called the IBM Automation foundation (IAF). IAF is the glue that binds together the IBM Automation offerings and provides the additive value as the Cloud Paks are used together. The following shared automation services are available in IAF:

- Embedded Watson for machine learning and natural language processing

- Event and data analytics cloud (Apache Kafka events, Flink streaming analytics)

- Process mining

- Robotic process automation (RPA)

IBM Automation is one way to explore many of the technologies that will be discussed in the chapters to come. So, if you are looking to getting "hands-on experience" with robotic process automation, intelligent document processing, integration, and APIs and AIOps, simply go to IBM Automation and give it a try.

UP NEXT

This chapter has outlined the basic definition of AI-powered automation, but we're really just getting started. In the chapters to follow, we further define and examine the process behind AI-powered automation by expanding the view to include a deeper dive into its key capabilities and architecture. We also explore a set of industry use cases that have put these technologies to work in meaningful ways.

The chapters to come cover similar topics explored in the episodes of our *The Art of Automation* podcast, where guests on the show are subject matter experts, sharing different aspects of AI-powered automation. These very same subject matter experts have been invited to co-author chapters in this book and share their examples of how automation is changing everyday life for the better.

Make sure you check out *The Art of Automation* podcast, especially Episode 0, in which Jerry kicks off the podcast series with a view of the "art of the possible "by elaborating on the key concepts of automation, self-driving cars, and how this all relates to the future of enterprise technology.

Chapter 1 – Figure 4. Podcast Episode 1 – Introduction to The Art of Automation

Chapter 2

ROBOTIC PROCESS AUTOMATION (RPA)

RPA is made up of three core technologies: workflow automation, screen scraping, and AI. The unique combination of these technologies allows RPA to solve the productivity challenge of manual desktop tasks.

Chapter Authors: Allen Chan and Jerry Cuomo

COVERED IN THIS CHAPTER

- Definition of RPA

- Best usage and intended users of RPA

- Practical limitations of RPA

- RPA and Artificial Intelligence

- What's next for RPA?

DEFINING ROBOTIC PROCESS AUTOMATION

Robotic process automation (RPA) is a program (in this case, a software robot) to mimic human users' interaction with their desktop to perform tasks (for example, copying information from an Excel spreadsheet to a form or inserting customer data and placing an order on a website). While we assume many human tasks have been automated in today's digital world, there is still a large portion of our daily work that requires manual labor, and much of that work is repetitive.

Imagine you are a data clerk responsible for processing incoming invoices sent to you by email or fax. You will have to read the incoming invoice—it could be a PDF document or a fax image—and enter the order manually into your ordering application. If this is a new customer making the order, you might also have to manually create the customer account. If you have RPA, the robot can leverage various optical character recognition (OCR) techniques and intelligent document-processing techniques to read the invoice and then simulate the mouse clicks and keyboard strokes on the computer screen to enter the information into the ordering application.

One key difference between RPA and other automation methods is that RPA is not limited to scripts, commands, or API but can also automate user interface interactions by recording and playing keystrokes and mouse clicks. Despite advances in various modernization techniques, there are still many legacy business applications (e.g., CICS, IMS, SAP) or native applications (e.g., Windows-based) that do not provide modern APIs or command line to automate. In some cases, the user just doesn't have access to the APIs (imagine you're using a third-party web-based application, like a banking website or an online bookstore), since the chances of the ITOps team giving regular users access to their backend API are very small. To automate tasks involving these systems, you need RPA.

THE THREE CORE TECHNOLOGIES IN RPA

Robotic process automation is made up of three core technologies: screen scraping, workflow automation, and artificial intelligence (AI). It is the unique combination of these three technologies that allows RPA to solve the productivity challenges that are hindering workers across all industries.

Screen scraping is the most basic feature of RPA involving capturing visual data from a computer screen and cross-checking it against reference information to decipher it.[16] While it is not the fastest or most accurate automation method, it proves useful in screen navigation and as a simple form of computer vision using optical character recognition (OCR).

When RPA was first introduced, there was an impression that it was just screen scraping technology. In a sense, that's not wrong; RPA is an evolution of screen scraping with a smarter use of variation technology, like screen assistance, more intelligent parsing of UI data (e.g., native Windows controls, Web browser DOM model), and a more scalable way of managing many robots at the same time.

Before the mass-market introduction of RPA, there were roughly three categories of *workflow automation*: fully manual, semi-automated human-centric, and fully automated straight-through process. The purpose of almost all automation projects is to shift the percentage of fully manual processes to fully automated straight-through processes. The desire to have as many straight-through processes as possible is what helped drive the API economy, since every service must be API accessible and programmed to eliminate all human interventions:

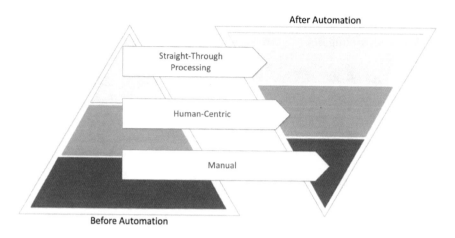

Chapter 2 – Figure 1 – Progression and value of automation

The challenge is that there is an investment needed between human-centric process and straight-through process, and sometimes the investment required can be significantly more than the benefits gained. As a result, we see that many initial automation projects are focused on business-critical processes where the ROI would be stronger. These sets of critical business processes often only account for the 10% of the processes in the company, and the majority of the rest are what we call "long-tail processes" and human-centric, but they are not significant enough to justify the investment required to build a new API or process reengineering.

This is where using RPA is compelling and practical. RPA is good at automating a set of repetitive desktop tasks that would otherwise be difficult and time-consuming to automate without proper API integration (see Chapter 8 – "APIs" for a more detailed explanation of APIs in the context of automation). The fact that many RPA solutions (including IBM Robotic Process Automation) provide low-code authoring tools combined with screen recording and smarter vision technology makes it even easier and faster for users to build the solution.

What this accomplishes is that we have a middle ground between human-centric and straight-through processes where we are using a hybrid approach involving humans, robots, and APIs to drive automation:

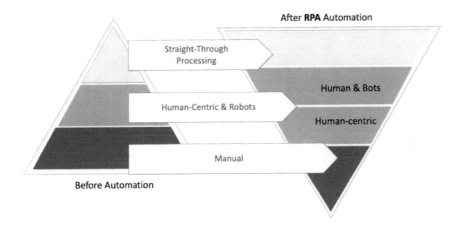

Chapter 2 – Figure 2 – RPA's hybrid approach involves humans, bots, and APIs

INTENDED USERS AND BOUNDARIES OF RPA

To identify if there are opportunities to use RPA within your organization, there are three places you can look into first:

1. Areas where you have a medium to large population of human task workers that are largely doing repetitive and manual work (e.g., order processing from emails or record reconciliation between systems).

2. Disparate systems that do not have APIs or where APIs are not accessible. Typically, we would have considered these situations as not automatable due to lack of APIs, but it is now possible with RPA.

3. Manual steps as part of a bigger task. I generally refer to these as micro-tasks. For example, as part of preparing a sales report, the

sales executive might need to copy groups of data from different marketing websites. While the sales strategy would still require the sales executive's intuition and experience, copying the data or formatting the report can be done more routinely using RPA.

ATTENDED VS. UNATTENDED RPA

There are two major forms of robots in RPA: attended and unattended. Attended RPA bots are ready and waiting to be activated by employees whenever they are needed to help the process along. Unattended RPA bots operate on a preset schedule, or as triggered by logic in the process flow. Attended bots can run on workstations or private servers or in the cloud.[17]

There are two main advantages to attended bots compared to unattended bots:

1. Attended bots allow users to automate a subset of tasks as part of the larger human-driven and more complex process where full automation might be difficult or wouldn't produce the best outcome (e.g., when certain knowledge-based decision-making has to take place in between the steps).

2. Attended bots allow users to run automations from their computers without requiring IT to provision additional computing resources.

Recently, we are seeing an increasing trend where companies are deploying more attended bots in addition to the more traditional unattended bots. This emphasizes the theme discussed in this book's introduction of the emergence of a hybrid workforce where automating actions in an enterprise becomes a "team sport."

LIMITATIONS OF RPA

- RPA—or robotic *process* automation—is a misnomer. It should have been called robotic *task* automation. RPA is quite good at automating a task with simple workflow automation, but it is not intended to be used to orchestrate work across multiple people or multiple systems. One would typically use business process management (BPM) software like IBM Business Automation Workflow for that purpose because BPM is more suitable for more complex interactions between multiple humans and automation.

- There are also many tasks that require human cognition and intuition. RPA bots are programs and can make use of AI to help them to make sense of the world, but they cannot think by themselves beyond simple and well-defined tasks. Some RPA vendors might lead you to believe RPA can solve all automation problems, but in reality, customers have overused RPA with unachievable expectations and are now realizing they need a more holistic end-to-end approach on their automation solutions.

- RPA does not replace API integration. In places where you have APIs and can use API-to-API integration, it is almost always more reliable and scalable, particularly where performance metrics and business analytics are also required. With that said, it is possible for an RPA bot to call an API. So, the two are not mutually exclusive.

USES OF RPA USED TOGETHER WITH OTHER AUTOMATION TECHNOLOGIES

- *RPA as part of an overall business or IT process*: In this case, the bots will be participants in the overall process and will work on tasks assigned to them. When a bot fails, it will log an error leaving the human operators to investigate the failure reasons, or even complete the rest of the task manually. By using RPA as part of an overall business process, the bot can then delegate or escalate any failures to its teammates or manager.

- *RPA as part of an integration solution*: When used together with an integration product like IBM App Connect, RPA can be part of an overall integration flow by basically providing an API to otherwise non-API systems.

- *RPA leveraging business rules and AI to make better decisions*: By using business rules to help the bot make better judgments while making the overall system more robust.

- *RPA leveraging intelligent document processing*: RPA can use intelligent document processing to read unstructured and semi-structured documents. Intelligent document processing has advanced significantly in recent years in terms of the variety of types of document it can handle accurately and also how it can be integrated into other automation tools and workflows. Using intelligent document processing as a built-in feature of RPA can help you automate even faster and simpler.

- *RPA as an interactive virtual agent, leveraging natural language processing (NLP)*: You can now leverage the power of conversation to initiate or provide additional information to the execution of bots, which helps incorporate RPA into your existing workflows, making it more consumable for non-technical users. Top RPA vendors (including IBM) embed NLP features in their standard RPA offering.

HOW AND WHERE ARTIFICIAL INTELLIGENCE WILL HELP

Today, there are several areas where RPA makes use of AI (with the full expectation to expand to more use cases in the future). Fundamentally, what RPA tries to do is to mimic humans' actions as they are performing their tasks.

These are four examples:

1. *Reading unstructured or semi-structured documents*: For example, invoices, scanned identity cards, handwritten notes, or email. In these cases, RPA can make use of basic OCR software, but for the robot to be smart, it will have to use a more advanced content extraction system like the IBM Automation Document Processing capabilities, where it uses deep-learning techniques to extract contextual information from invoices or id cards (e.g., getting the customer address and item part numbers or even performing automated correction of information).

2. *Integration with chat and voice*: A typical example is using RPA to build chatbots or voice bots—integrating with existing services like IBM Watson Assistant. IBM Watson Assistant uses AI to identify and create recommendations, spot trends and emerging issues as they happen, and automatically learns from user choices to improve interactions.

3. *Computer vision*: Reading the screen using computer vision to understand the user interface widgets (e.g., Windows, scrollbar, button) instead of relying on explicit screen pixel coordinates. This helps provide a more accurate and reliable automation and makes bots more portable to different devices.

4. *Automating tasks discovered by task mining*: This is a relatively new area that was introduced with the idea that we can identify and create highly effective bots by observing how people work. This also gets into the concept of "learning" bots. This is obviously very desirable

and has huge ROI benefits. IBM sees this as an active area of research for RPA and automation in general.

ROBOTIC PROCESS AUTOMATION IN ACTION

This section provides a quick and simple demonstration of RPA in action using IBM's robotic process automation tools.[18] In this example, you will witness how Bill, a business analyst for Focus Corp, uses RPA to build, run, and monitor a software bot to automate a manual task.

Bill is looking to automate the task of a customer-initiated change of address. This task is performed by the customer and is usually done from the Focus Corp website. As the figure below illustrates, when such a request is made, a Focus Corp employee manually enters the address change in Focus's master account management system (MAMS). Focus's MAMS is a legacy system and does not have published APIs, and hence the address change request cannot be automated through programmatic means. Manually entering address changes is repetitive, time-consuming, error-prone, and of low value to the team's time. With RPA, Bill can build a software bot to automate this work by emulating a human performing data entry. Once Bill's team is freed of this mundane task, it will allow them to focus on higher value work like onboarding new customers and trouble-shooting operational issues.

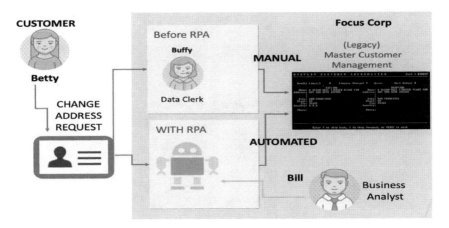

Chapter 2 – Figure 3 – Focus Corp's
process of address change with and without RPA

Bill starts building his bot in RPA Studio, a low-code, AI-driven bot authoring environment. With Studio he instantly has access to over six hundred prebuilt commands to quickly assemble his bot using guided wizards. Utility commands are available for a variety of tasks, including commands to automate the manipulation of email files, Microsoft office files, and data found in web browsers. More advanced commands use AI, including machine learning and natural language processing, for content extraction, document classification, chatbots, and much more.

As illustrated in the following figure, Focus Corp stores pending customer address change requests in a Microsoft Excel file. Bill programs the bot by teaching it through a command to open the Excel file and extract customer names.

RPA Script

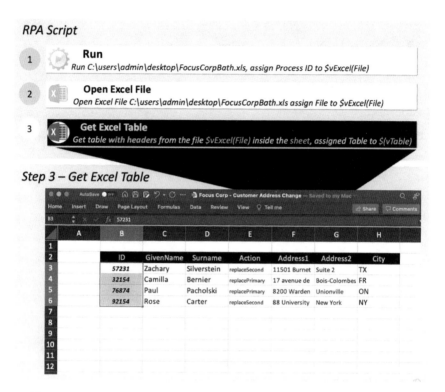

1 Run
Run C:\users\admin\desktop\FocusCorpBath.xls, assign Process ID to $vExcel(File)

2 Open Excel File
Open Excel File C:\users\admin\desktop\FocusCorpBath.xls assign File to $vExcel(File)

3 Get Excel Table
Get table with headers from the file $vExcel(File) inside the sheet, assigned Table to $(vTable)

Step 3 – Get Excel Table

	ID	GivenName	Surname	Action	Address1	Address2	City
	57231	Zachary	Silverstein	replaceSecond	11501 Burnet	Suite 2	TX
	32154	Camilla	Bernier	replacePrimary	17 avenue de	Bois-Colombes	FR
	76874	Paul	Pacholski	replacePrimary	8200 Warden	Unionville	ON
	92154	Rose	Carter	replaceSecond	88 University	New York	NY

Chapter 2 – Figure 4 – Extracting customer address fields from an Excel with IBM RPA

Now that the bot has access to the change requests, it needs to log into Focus Corp's legacy MAMS. The bot can be taught how to access usernames and passwords from a secure vault, to log into the MAMS. Once logged in, the bot can be further programmed to search for the first customer in the change list. Bill trains the bot using the Studio tool by watching and recording Bill's keystrokes.

If the search is successful, the bot will be programmed to update the customer address entry in the MAMS with the new one found in Excel. This is repeated for all customer requests until all are complete.

In minutes, Bill created and fully trained a bot to perform the change-request task. When called upon, the bot can rapidly iterate through the customer request changes, which often involve updating over twenty data fields in the MAMS, in under thirty seconds. Now that bots are busy performing this repetitive task, Bill can monitor the team's robots via the IBM RPA tools dashboard. The tool provides out-of-the-box views that can be customized to display the data needed to monitor his robotic workforce. The following figure includes metrics like execution count, box status, and day-of-week views.

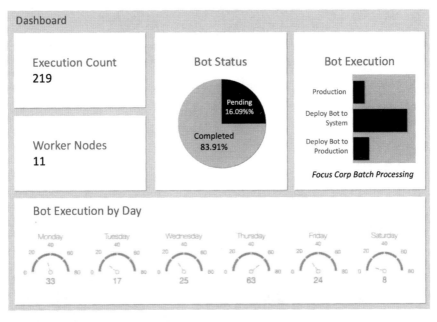

Chapter 2 – Figure 5 – IBM RPA dashboard displaying bot metrics

In fact, Bill notices a pattern developing on the day-of-week view and decides to add capacity to the bots on Thursday. This is made easier with the RPA tools workload management (WLM), which distributes work across multiple bot agents. The WLM feature can be used to efficiently manage the cost of running bots by allowing bots to share the same virtualized hardware resource.

In summary, Bill built, ran, and monitored an untended bot to automate the input of customer information into legacy systems to leave more time for his team to perform higher value customer interactions. The RPA bot saves employee time, reduces human error, speeds access to update customer information, and reduces total cost of ownership.

WHAT'S NEXT FOR ROBOTIC PROCESS AUTOMATION?

RPA's undeniable strength is in clear ROI and being programmable by everyday business users. RPA, as a concept, is very powerful and possesses great potential for further innovation and application. RPA allows enterprises to bridge from their legacy or existing systems into the more modern API economy without requiring them to modernize their platforms first. There are several areas where RPA will evolve:

- The robots will get smarter and understand the world better by leveraging different AI techniques like natural language understanding, computer vision, and document extraction. This allows RPA bots to help with more tasks.

- Everyone will have a bot on their desktop. We see companies deploying bots on every employee's desktop to help everyone in the organization to automate part of their everyday work.

- RPA being more directly integrated into process automation software including workflow management systems and process-mining software. Process automation software brings more complete visibility into business processes which will better identify opportunities where RPA bots can be used.

Creation of the bot will get much easier than today. There will be better recording capabilities, more intrinsic use of computer vision, and leveraging of task-mining technique to create bots.

UP NEXT

To learn more about RPA, make sure you check out *The Art of Automation* podcast, especially Episode 4, in which Jerry and Allen Chan discuss how RPA automates repetitive human tasks by mimicking human actions and how this can save time in tasks such as reading documents and interacting with applications. Allen describes how robotic process automation is a misnomer because it's not the entire process that's automated but the tasks within it. He makes sure to clarify that RPA is a tool built to assist humans, not steal our jobs.

* * *

Now, there was a bit of a debate trying to decide which chapter to start this book off with. While RPA ultimately won that debate, there was a good argument to start the book off with process mining. As you will see in the chapter that follows, process mining and RPA are like "peanut butter and chocolate." They go very well together because process mining is often the tool that determines the best places to apply RPA.

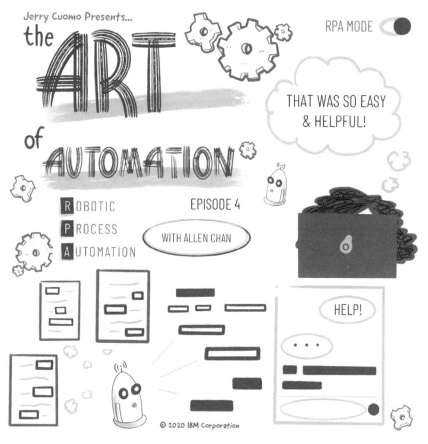

Chapter 2 – Figure 6 – Podcast Episode 4 – Automation with RPA

Chapter 3

PROCESS MINING

Process mining gives you the tools and methodologies that you need to unlock the data that shows how processes work, how people do their jobs, and where problems are coming from.

Chapter Author: Harley Davis

COVERED IN THIS CHAPTER

- What is process mining?

- Process-related data collection

- Process analytics

- Taking action

MAKING A CASE FOR PROCESS MINING

Imagine this:

- You are the back-office manager in a bank, responsible for the teams that open, manage, and close customers' accounts. Your service level agreements (SLAs) have been sliding for several months, with times for opening and closing accounts outside regulatory boundaries and customer complaints piling up. You need to figure out what is happening and do something about it quickly.

- You are the CIO in a major corporation, with a large IT help desk team responsible for ensuring that tickets are dispatched and solved expeditiously. The number of tickets keeps growing as your company uses an ever-growing number of complex applications, but your budget is stable. The net promoter score (NPS) for the help desk is going down. How do you make sure you can keep up with the growth with a team that stays the same size? Where do the issues come from that are causing delays?

- You are responsible for a purchasing and accounts payable department. Despite all your efforts, you keep hearing reports of maverick buying outside company guidelines. And many of your suppliers are being paid late and the penalties are starting to take a toll. Where do the problems come from? How can you solve them systematically?

For all these problems, it seems like increased automation could help. After all, you can't ask your teams to keep doing more in the same amount of time, and more streamlined automated processing could solve some of the issues. Automation can handle the routine issues and thus help free up peoples' time to work on harder issues and improve customer service.

Maybe you have brought in external consultants who have told you about the wonders of robotic process automation (RPA), AI-powered decision management, and business process management—and those things seem to work elsewhere. Maybe you have used RPA on a specific task, but it did not improve the overall process. Your business analysts have drawn up process maps and interviewed employees and have some ideas on where the issues stem from. But the investment in automation needs a clear business case for your management to agree to spend money ahead of seeing results, and so far, the business case is a bit wishy-washy. You are not sure the ROI will really be there.

What is missing in this scenario?

In all the examples above, management has a good intuition that something needs to be fixed, and the KPIs used to measure the business are showing there is a problem. But discovering the true roots of the problem and ensuring that proposed solutions will have the right impact require something else—actual data about how business processes are being executed today, root cause analysis of this data, and simulation of the impact of automations on an accurate model of how the business works.

Process mining[19] brings that missing link to the table.

Process mining gives you the tools and methodology that you need to unlock the data that shows how processes work, how people do their job, and where problems are coming from. It gives you analytics to dig deeper into the business and uncover where automation and other process changes can have the biggest impact. It can then simulate how the business would operate with the automations in place, letting you focus on the solutions that maximize expected results, radically increasing your confidence that your investment in automation will really pay off. Then you can use the same tools to measure the improvement against the estimates and help you continue your journey to automation.

In this chapter, we will describe the various steps in process mining and some of the analyses available to help you create your own data-driven road map to automation. We will be using IBM Process Mining, the IBM-enhanced version of myInvenio, as the tool of choice. IBM Process Mining has an especially rich set of analytics and simulation capabilities, with links to the rest of the IBM Automation portfolio. It includes capabilities like business rule mining, task mining, multilevel process mining, reference model comparisons, and the ability to create simulation data for a process model to get insights independently of historical data.

WHAT IS PROCESS MINING?

The overall paradigm of process mining is straightforward, as illustrated in the following figure.

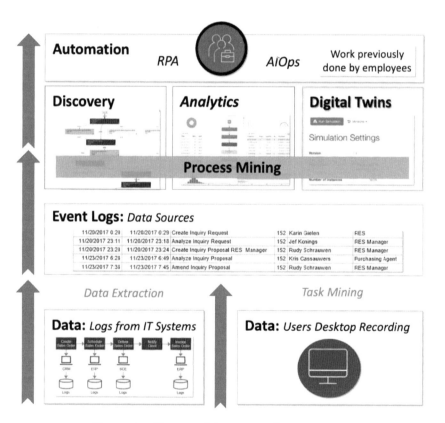

Chapter 3 – Figure 1 – The process of process mining

The idea is to go from process execution data (found in either system logs or recorded from peoples' desktops), to analyses of that data to help understand how the process works, to discovering where there are meaningful opportunities for automation, to simulating the impact of the proposed changes using the model created during the analysis. Then you can create the automations that will have the most impact using RPA, decision management, and the other technologies in your automation toolbox. To close the loop, you can then measure the impact of the changes by gathering new data from the updated process and repeating the cycle again.

Let's look at each of these areas in a little more detail, then dive into some of the analyses and tools you can use to dig into your business and IT processes.

DATA COLLECTION

The first step in doing process mining is to gather the data that will be needed for the analysis. This is usually the most time-consuming part of a process-mining project. You will need to figure out where the data sits, how to access it, and how to format it in a way that the process-mining tool can use.

We are looking for data that shows how people execute processes. There are two primary sources of this data: system logs from systems that people interact with and records of the actions that people do on their desktops when engaged in executing a process.

System logs

Systems that people use include ERP systems, CRM systems, issue ticketing systems, accounting systems, and so on. You first need to do an inventory of the systems people use in the process you are trying to analyze and the kinds of information these systems store. Often, these systems put an entry in a log or database whenever someone executes a transaction or change with them. We are looking for these transaction or event logs. The data should include the action executed (which task was being done), an ID of a process being executed (typically a contract number, client number, ticket number, or similar), a timestamp for when the event occurred, who executed the action (the user ID), and maybe other information that is interesting for the analysis—how long something took, the outcome of the event, and other interesting information.

This phase of the project needs the involvement of the folks in IT who understand how these systems work, where the data can be accessed, and what its format is so that it can be read and transformed into the right format for the process-mining tool. For some systems, such as SAP and others, the process-mining tool helps by providing predefined connectors that do a lot of this work.

Task mining

The other main source of useful data is watching what people do on their desktops when executing processes. We can install recorders on the desktops and configure the recorder to store events whenever a user does something related to their process execution job (and ignore everything else they do to ensure privacy for unrelated activities). Then, we can send these event logs to a central server where they are consolidated with all the other peoples' records; when enough data has been gathered, it can then be fed to the process-mining tool for analysis.

System logs and task mining are complementary ways of getting historical process execution data and are often used for different purposes. System logs are good for doing an overall process analysis and seeing the big picture, especially for processes that are centered on modifying data in one system or a small number of systems (e.g., ERP systems for accounting, CRM systems for sales and marketing processes, or IT ticketing systems for help desks). Task mining is good for getting down to the details— exactly what actions does a person take to execute a task, under what circumstances, with what variations, and so on. This is very helpful when you are considering automating these actions using RPA tools that focus on that fine-grained detail.

ANALYTICS

Once the event data has been prepared and fed into the process-mining tool, it can analyze the data to produce a set of visualizations that can be used to pinpoint problems.

One basic analytics visualization is the process map, which shows the set of tasks executed during the process and how they are connected—which ones follow which other ones, the order in which they are executed, etc. Because the process may follow a different sequence depending on the different types of cases, the process map shows the different "process variants" that have been used. In addition to showing the process map, the analytics can show which tasks and which variants are executed most often, which ones take the most time or which ones cost the most. This is your first clue to finding issues—the tasks and sequence variants that are seen most often that take the most time or that cost the most are good candidates for further investigation.

The figure below shows how this works. Each task in the process is a labeled box. The darker the color of the box, the more often it is executed in the data set provided. The number in the box shows the number of times the task was executed. For example, in the following figure, task "Authorization Requested" was executed 46,415 times. The arrows indicate which tasks follow which others. In our case, the "Authorization Requested" task led to the "BO Service Closure" task 44,560 times (presumably, the remaining 1,855 cases were either rejected or still pending when the data was analyzed):

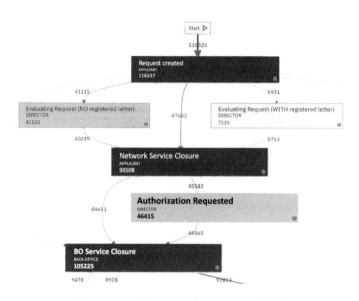

Chapter 3 – Figure 2 – Process Map: Tasks

This starts to become even more interesting when we look at the time spent between various tasks in the figure that follows.

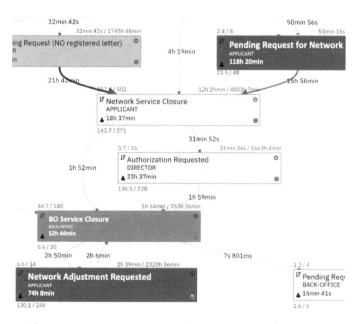

Chapter 3 – Figure 3 – Process Map: Time spent during tasks

The darker colors indicate that more time is spent on that task. We can see a couple that pop out as problematic—"Pending Request for Network Information" and "Network Adjustment Requested" seem to take a lot of time. Perhaps there is something we can do to automate the network information and network adjustment requests that would speed them up? You can see how powerful this kind of information is.

You can also see, in the following figure, which paths through the tasks are happening with what frequency, which is another clue to finding issues.

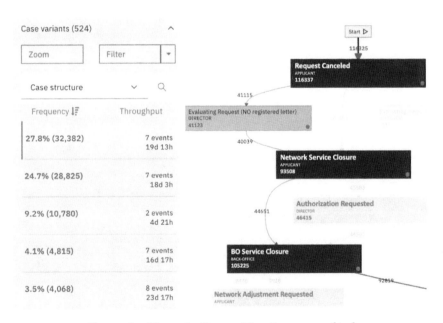

Chapter 3 – Figure 4 – Process Map: Frequency of task

In this example, the highlighted path is the most frequent way this process is executed—used 27.8% of the time and taking nineteen days and thirteen hours, on the average. If we can improve this variant first, it will probably have the biggest impact on the overall process. In this variant, there is no Network Information Requested or Network Adjustment Requested, so maybe our previous idea was not, in fact, the right place to look. If those tasks are not executed very often, improving them won't help the overall KPIs, even if we improve them by a lot.

But how can we see these KPIs? As you can see in the figure that follows, the tool allows us to compute KPIs based on the data available, and how those KPIs respond to the different areas at which we are looking.

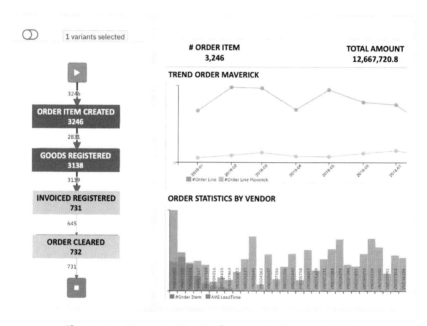

Chapter 3 – Figure 5 – Key Performance Indicators (KPIs)

Here, we can see how a KPI (namely, maverick buying) can be displayed in the context of a process variant. In this example, we can see both the amount of maverick buying over time and a breakdown of maverick buying by vendor. This could point us to some products and vendors to focus on reducing maverick buying, which might lead us to introduce automatic alerts or process redirects when purchase orders for those vendors are detected.

Another clue to finding process irregularities comes from conformance checking—finding process variants that do not correspond to a predefined process map that shows how the process should be working. In IBM Process Mining, you can upload a "reference model" that indicates the prescribed way a business process should be executed, for instance, by using a process map made with BlueWorks Live. The analytics can then compare that reference model to the actual data and point out inconsistencies. These are typically good places to start looking for errors and problems.

The image below shows how the conformance checking is displayed in the tool. In this example, we can see that 39,300 cases are non-conformant, and the red tasks indicate which ones are unexpected for particular process variants. In this case, we can also see in the following figure that the tool has calculated that $3,439 is spent per non-conformant case, based on time spent and the cost of people's time.

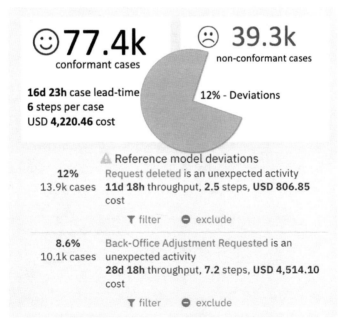

Chapter 3 – Figure 6 – Conformance checking

The tool has many other ways to display information and drill down into the data. A couple of these include the following:

- *Business rule induction to indicate how the process moves from task to task*: The tool can figure out the business rules that determine how the process moves along. These insights are valuable for finding the set of circumstances that determine how your business takes process decisions and can be the basis for automated process decisions using IBM Automation Decision Services (and improved decision-making overall).

- *Role-based analysis*: The tool can slice the data according to roles, helping you understand how different people in different roles execute their parts of the process differently and where there are bottlenecks. This can indicate where particular departments might have issues and help you adjust staffing and expertise levels in your organization.

There are many more. As you gain expertise with the tool, you will discover many new ways of gaining insight into your business.

TAKING ACTION

Once you have understood your process better, it is time to figure out what to do to improve the process. There are many things you can do, but let's focus on one area that is particularly useful with process mining—how to use RPA bots to automate bottleneck tasks to improve the overall process flow.

RPA bots, for the most part, replicate repetitive human actions on the desktop so they can be done more easily, thus freeing up people to spend their time on activities that require deeper thinking. Good candidates for RPA bots are tasks that are done often, that are repetitive, where you can save time by automating them, and that have an overall positive impact on the business KPIs (for instance, overall process resolution time that can lead to better customer service or alignment with regulatory deadlines).

Once you create an RPA bot, you can replicate it and execute it on different virtual desktops, and you can mix human activity with automated activity, depending on how much you want to scale the RPA solution.

When you use task mining with IBM Process Mining, as the following figure illustrates, the tool does a lot of the work for you in helping determine the best task candidates for RPA.

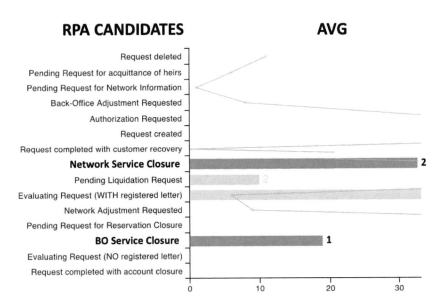

Chapter 3 – Figure 7 – Best task candidate for RPA

Here, the process-mining tool has pointed out two activities whose automation with a bot might have a big positive impact on the overall process: Network Service Closure and BO Service Closure. Furthermore, there are parameters that can be adjusted to estimate the overall impact depending on what percentage of these tasks are automated with a bot, and how many variant versions of these tasks are included in the bot. With the current parameters, the estimated savings per process instance is $369.36, and the overall savings in terms of human labor available for other activities is $43,054.76 for the 116,566 cases in this data set. These are definitely viable candidates for effective automation.

Now, we can also simulate the overall process execution in this RPA scenario, as illustrated in the following figure.

Process overview

	A	B
Case count	116,566	1,000
Average case lead time	18d 16h	16d 6h
Average case cost	USD 3,957.07	USD 4,046.40
Total case cost	USD 461,260,200.00	USD 4,046,400.00

Case duration and count

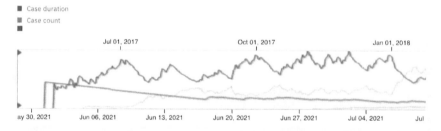

Chapter 3 – Figure 8 – Process simulation

Here, we have run a simulation with a certain percentage of human tasks in the original data replaced with the RPA bots that execute the tasks much more quickly. We can see that in this simulation run, the overall average process execution time for this account closure example went down from eighteen days and sixteen hours to sixteen days and six hours—a 10% overall time saving that has direct impact on customer satisfaction and regulatory compliance. The ROI for automation is clear and well-defined in terms of both financial savings and KPI improvement.

The next steps are to create and deploy the RPA bot, and then start investigating the real impact of the change using the same process-mining tool and setup. This will let us both determine if the ROI was actually met and point out the next set of automations to improve the business process. This repeated process monitoring, measurement, and automation is the motor that will drive ongoing business improvement, and that is one of the main drivers behind the excitement around process mining as the next technology advance in automation.

MOVING FORWARD

We have seen how we can take business process execution data, create detailed analytics for drill-down to understand how the business really operates, and then use those analytics to create automations that drive significant improvements to our business, both in terms of cost savings and customer satisfaction improvements.

But we have really just scratched the surface of what we can do with process mining. Here are some other areas where process mining can help:

- We looked at an example from banking operations with account management, but the potential of process mining is much vaster. It can be used for all processes with human activities. IT management is a typical example—managing the processes around help desk tickets, finding categories of tickets that can be better automated, finding manual steps in IT systems management that will make a real impact on operations. Process mining can be used to identify points of potential integration between different systems by analyzing where data gets held up today. It can be used in software development to identify inefficiencies in development, test, and deployment of software.

- We focused on RPA bots as an automation—process mining can also identify other types of automation. A few examples:

  Decisions: Often, processes are inefficient because we rely on people to make decisions that could be automated using a combination of machine learning and business rules. We can see in the process-mining analytics where there are loops in a process—tasks that are redone multiple times due to human error. We can spot approvals and other decisions that are taking a long time. These are all signs that automated decisions may improve the overall process.

  Business process management: Many processes would benefit from a better-coordinated and choreographed business process, with clear tracking of progress, task inboxes for employees, and flexible but coordinated case management. Process mining can create the outlines for implementing BPM—the process maps generated by the analytics engine can be exported as BPMN diagrams that serve as the basis for a business process management project.

  System-level integration: RPA is great for integrating systems that have only a UI (and no underlying APIs) and for simulating human interaction on a desktop. But a deeper integration of systems using their APIs and exchanging data directly with no human intervention at all is even better, when it is possible. Fortunately, process mining can also pinpoint where such system-level integration would make a difference and can lead to more successful projects.

- We started from available data to understand process behavior, but it is also possible to use IBM Process Mining's powerful simulation engine, along with the ability to import process models to do full process simulation in the absence of any actual data. By importing a model from, for example, IBM BlueWorks Live and providing some basic information about anticipated task execution times, IBM Process Mining can estimate how processes are likely to behave and where bottlenecks will occur, leading to opportunities for process optimization prior to putting the processes into production in your organization.

You can now understand why process mining is generating such excitement in the market and how it is really the best first step in your journey to automating your business.

UP NEXT

Make sure you check out *The Art of Automation* podcast, especially Episode 17, from which this chapter came. In this episode, Jerry and Harley Davis set out to answer the most frequently asked question by *The Art of Automation*'s audience: Where do we start? Harley explains the concept of process mining and how it takes a data-driven approach to translating an enterprise's desire for improved operations into actions and automations that can actually be implemented. He also discusses how process mining technology allows a business to simulate (using a "Digital Twin") what would improve if certain automations were applied, removing risk and justifying AI investments.

* * *

The initial chapters of this book cover two of the most fundamental and established topics in business automation: RPA and process mining. The next chapter introduces the topic of digital employees, an exciting new area of business automation that builds on these technologies and boldly extends them to something akin to Amazon Alexa meets J.A.R.V.I.S., the intelligent assistant created by Tony Stark from Marvel's *Iron Man* series.

Chapter 3 – Figure 9 – Podcast Episode 17 – Automation with process mining

Chapter 4
DIGITAL EMPLOYEES

Augment your team, automate day-to-day work, and elevate employees to focus on what matters most. Digital employees are the way of the future. We know it. When and how is in our hands.

Chapter Author: Salman Sheikh

EMPLOYEE – A WORKER EMPLOYED BY AN EMPLOYER

While there are many excellent books on digital workers and chatbots,[20] for this *Art of Automation* book, we've decided to focus less on those topics and highlight an exciting new technology that provides capabilities that extend the best of chatbots, RPA, and digital workers. Enter *Digital Employees*.

In the fall of 2020, Allen Chan, CTO of IBM Business Automation, was roaming the virtual halls at IBM socializing the idea of Digital Workers 2.0 to fuel the future of work. In parallel, I was tasked with bringing to market a new interactive AI technology coming out of IBM Research labs, codenamed "Verdi." Verdi was an AI-based planning algorithm that intelligently orchestrated agents based on natural language instructions.

While this was happening inside IBM, the automation market was undergoing exponential growth in RPA and Gartner's hyper-automation concept was inspiring thoughts of an integrated automation platform. In parallel, the conversational AI trend had also reached a level of adoption. Conversational AIs, or chatbots, were now commonplace in help desk and customer support scenarios, offering a rich natural language interface to answer pre-trained questions and execute predetermined actions. The next evolution of Conversational AI, predicted by Gartner, was to be multimodal, with deep context and with a wide range of enterprise app integrations to tackle more complex virtual assistant use cases.

These market trends did not wholistically capture the concept of a digital employee, but the technology landscape offered a level of maturity to set the foundation for IBM's point of view on a hybrid workforce. IBM's integrated automation platform (Cloud Pak for Business Automation), IBM Research's interactive AI innovation (Verdi), and Allen Chan's Digital Worker concept came together to form Watson Orchestrate. As announced by IBM's CEO, Arvind Krishna, at IBM Think 2020, Watson Orchestrate empowers companies to compose their own digital employees to augment their human workforce and automate work when and where they need it.

WHAT IS A DIGITAL EMPLOYEE?

The best way to understand what a digital employee is, is to focus on the "employee" part first and then the "digital." Meaning, let's understand the job role and function of the digital employee and then focus on the technology that makes it possible. Critical to understanding the job role and function of a digital employee is to first understand the job role and function of the human employee it is going to support.

Let's consider a real example:

Digital Employees in Human Resources

Meet Pat

Pat is an HR business partner in a large enterprise. She is responsible for working with business units within her company to plan and execute strategic human resource goals. One of those goals is to ensure high employee engagement and retention through an effective employee promotion cycle.

Pat and her team have a quarterly promotional cycle. Each quarter, the criteria for nomination are finalized in partnership with business leadership. Then the nomination process is kicked off with more than a hundred managers to submit their nominations back to HR. Once nominations are received, HR needs to aggregate the nominees into a master list, augment each nominee record with 120+ data points, and then submit the nominees for approval to both business and HR leadership.

Meet HiRo

HiRo is a digital HR business partner and Pat's new sidekick. Pat is still responsible for the overall quarterly promotional cycle, but HiRo is here to do a lot of the legwork so that Pat can focus on more strategic conversations related to the promotion cycle with her business stakeholders.

HiRo interfaces with Pat and other employees using Slack, the chat platform deployed at Pat's organization. That is, Pat, and other employees, can chat with HiRo using Slack.

SKILLS

Kickoff Email
Kickoff quarterly nomination process by notifying managers.

Add to Nomination List
Submit employee as nominee for promotion.

Nomination Approval
Launch the executive approval workflow.

Workday
Get and update employee records.

ADP
Access payroll and salary information.

HiRO

Digital Business Partner
IBM CHQ, HR Services
IBM Human Resources

 Hiro@us.ibm.com

 @Hiro

TEAM

Reports to:
Patricia

Peers:

HiRO1

Jeri (Talent Partner)

Consultants

JOB-SPECIFIC EXPERTISE

Digital HR Partner
Responsible to work closely with Human HR BP to proactively address business issues related to workforce & productivity.

Data Specialist
Expertise in making available, integrating, and optimizing structured and/or unstructured employee data.

Chapter 4 – Figure 1 – HiRo's digital employee profile

HiRo has been built with key HR skills:

1. *Kickoff Email*: Every quarter, Pat tells HiRo to fire off the kickoff email to the hundred-plus managers. HiRo gets the latest promotion criteria as well as the due date for the nominations from Pat. Taking this information, HiRo populates a predefined email template and fires it off to the managers letting them know when their nominations are due.

2. *Add to Nomination List*: When a manager selects an employee to nominate, HiRo adds that employee to the nomination list.

3. *Nomination List Approval*: Once all the nominations have been received, Pat tells HiRo to generate the Master Nomination spreadsheet. This skill not only merges the over-one-hundred nomination lists received from the managers but also augments the employee data by pulling more than 120 data points from various HR systems (performance management, compensation, etc.).

HiRo also has skills to help employees perform key tasks in the popular business applications, including Workday (for human resources) and ADP (for payroll). But just looking at the three skills above, HiRo is already saving Pat and every HR business partner in her team days of chasing managers, pulling and validating data, and crunching spreadsheets. HiRo is able to pull employee data for each nominee from multiple HR systems in seconds, a task that would take HR days to complete.

ANATOMY OF A DIGITAL EMPLOYEE

A digital employee is a complex set of technologies that need to come together to form a whole.

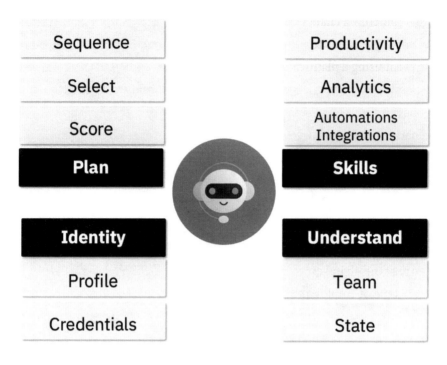

Chapter 4 – Figure 2 – The anatomy of a digital employee

Planning – the key to intelligent orchestration

There are plenty of existing automation technologies that provide orchestration of work. Companies have deployed workflow and case management systems to orchestrate end-to-end business processes, such as processing applications, claims, support tickets, invoices, etc. The orchestration provides a standardized and, ideally, optimized workflow. This workflow however is predefined, and each decision point and path are predetermined. In real life, things change and don't always fit into predetermined scenarios. That's where intelligent orchestration can fill the gap.

The key to intelligent orchestration is *planning*. A digital employee interfaces with the world around it through *events*. It interfaces with human employees through conversation, and this triggers natural language events.

It can also interface with business systems by listening to notifications when something changes; this triggers data events. Given an event, the digital employee needs to find the best course of action to address the incoming event using a planner. The planner evaluates the set of skills it has at its disposal and scores them based on a confidence-rating that estimates the probability of the skills ability to address the incoming event. In a simple scenario, a manager asks HiRo to pull an employee record. HiRo has a Workday skill to "get" an employee record. This skill scores at >90%, letting the planner know it should be selected to handle the incoming event. But what if HiRo also has an ADP skill to "get" employee payroll information (i.e., another type of employee record)? This ADP skill would also score high. This is where the intelligence of the planner and conversational nature of the digital employee enable it to go back to the manager to clarify which type of employee record they need. The planner can also learn from this interaction and begin to weigh one skill over another to help it optimize its selection algorithm for the future.

So the planner can understand the incoming event and select the best skill. It can also disambiguate between similar skills. Additionally, it can learn and improve its scoring algorithm over time.

But this is just the beginning for intelligent orchestration. The planner also needs to be intelligent enough to sequence skills as needed to truly orchestrate a workflow on the fly. Once HiRo retrieves the employee record for the manager, the manager decides to add the employee to the promotion nomination list. The nomination list will be reviewed by upline management, who needs key data points not available in the core employee record to decide if the employee nomination should be approved. Data points such as time in current position, last salary increase, etc. HiRo has the skills to retrieve this information from the various backend HR systems. When the manager asks HiRo to add the employee to the nomination list, the "Add to

Nomination List" skill lets the planner know that it requires the following input parameters to execute the skill:

- Core employee record

- Time in current position

- Last salary increase amount

- Last salary increase date

The planner already has the core employee record but is missing the other parameters. It does have the ADP skill to get employee salary information, however. Here's where the intelligence of the planning algorithm kicks into overdrive. It knows it needs to execute the "Add to Nomination List" skill, but it can sequence the ADP skill ahead of it in execution to get the missing input parameters. The planner does this by looking at all the skills it has at its disposal and determines which ones can output the values needed as input to the "Add to Nomination List" skill. Using this technique, you can see how the planner can sequence two skills, but it can also be extended to sequence three or more skills. This is true intelligent orchestration.

Understanding – adds context to intelligence

Going hand in hand with planning is understanding. Understanding brings context to the intelligence. This gives the intelligence an understanding of the state of the world as it knows it today.

It starts with foundationally knowing the self and its surroundings. For digital employees to function in an organization, they need to know where they fit into an organization, which team they belong to, and whom they work for. They have direct knowledge of the user directory and have their own record in the org tree. This gives the understanding of which employees they work with and which manager they report to, so they have a clear path of escalation when they need help. And digital employees do need help from time to time.

The second layer of understanding is knowing what objects the digital employee can work with. Each digital employee maintains a schema describing the objects it knows how to work with. In the example above, HiRo is built with the understanding that it will work with employee record objects. When HiRo retrieves the employee record for the manager, it can add it to its memory of objects. This enables the manager to work with HiRo over multiple interactions using that object. The manager can now ask HiRo to get more information on this employee, for example last year's performance review. The manager can also tell HiRo to add this employee to the promotion nomination list. In either case, HiRo understands the context (i.e., the employee) of the conversation and completes the requests on behalf of the manager.

Identity – gives purpose

Much like human employees, a digital employee exists to satisfy a specific job function—it has a specific purpose. Each digital employee has its own employee profile that personifies its purpose. For example, HiRo is a digital HR business partner. When employees want to interact with it, they know what its purpose is, what it can do for them (i.e., the skills), and how they can interact with it (i.e., via its chat handle, email address, etc.).

With its identity, it also has its own credentials to the various business systems it needs to work with. HiRo has access to Workday but has its own credentials so HiRo's manager can control what actions and data HiRo can work with in Workday. If HiRo is meant to serve only the North America team, HiRo's credentials can be limited in Workday to work with only the NA employee data. These credentials not only give managers a way to control access for digital employees as they would their human employees but also give all employees a way to track what the digital employee has done in their business systems.

Skills – get work done

Now given all the intelligence and identity of the digital employee, though fairly smart, it is not very useful unless it has the ability to execute work. That's where skills come in. Skills are the atomic unit of work that a digital employee can perform.

Skills come in a variety of flavors:

1. *Productivity Skills*: These enable the digital employee to work with common productivity tools to send emails and schedule events. Imagine if a manager asks HiRo to pull an employee record, they can easily ask HiRo to also email that record to a colleague for review. That's convenient!

2. *Analytical Skills*: Employees spend countless hours pulling data from various business systems to generate reports and glean insights. Many others don't, even though they should, because its time-consuming or they lack the technical skills. This category of skills enables employees to get the digital employee to query business data using natural language. The digital employee generates the SQL statement behind the scenes and pulls the tabular results back as a chat response. And just as simply the employee can ask the digital employee to graph that data, and the digital employee will convert it to a graph. That's convenient!

3. *Automation Skills*: There is a plethora of efficient automation technologies. Whether it's orchestrating workflows, modeling business decisions, processing documents, or even automating employee tasks using RPA, the digital employee doesn't seek to reinvent these

automations. Instead, it aims to leverage them and make them accessible to employees. Imagine an army of RPA bots that are now accessible to your employees when and where they need them through natural language. That's convenient!

4. *Integration Skills*: Employees spend a good share of their time working inside of business applications, such as Salesforce, Workday, etc. This work takes them away from higher value tasks like having conversations with customers/employees or solving problems and making decisions. The digital employee is their sidekick that can take on the more tedious task of updating records in those business applications through integration skills. Integration skills leverage existing APIs to let the digital employee perform work. Using its unique credentials, it will only perform work and access data that it is allowed to work with. That's convenient *and* secure!

Bringing it all together

To summarize, a digital employee is the coming together of a complex set of technologies building on natural language understanding, an integrated automation platform, and an integration platform. However, a few additional groundbreaking innovations where needed to make digital employees come to life, such as the Verdi technology with its built-in planning (i.e., orchestration) and understanding (i.e., skill matching, scoring, selecting) capabilities. Brought together to empower businesses to compose digital employees to augment any part of their human workforce. The example we discussed here was a human resources use case, but Watson Orchestrate is already being envisioned to augment sales teams to automate work in salesforce, in insurance to help speed up the claims-approval process, and in finance to help relationship managers be better prepared for customer meetings. Digital employees help automate focused use cases for employees. Build the right skills and digital employees can automate any line of business. Fortunately for Watson Orchestrate, building skills is a low-code exercise that a line of business can do for themselves.

DEBUNKING COMMON MYTHS

Will digital employees replace the human workforce?

Digital employees and AI in a broader sense are just the latest technology enablers. Just like when computers entered the workforce, computers didn't replace workers, but they elevated the employees to be more productive. For example, before a cash machine, people had to handwrite receipts and hand-calculate totals. Letting a machine take care of that for them didn't replace them; it enabled them to do other more meaningful work. A digital employee aims to do the same for today's employees—take on the mundane or repetitive work to increase employee productivity, employee impact, and job satisfaction.

Digital employees are the same as conversational bots

Digital employees focus on automation; it's about building automations and making them available to your employees so they can use them on the fly to get work done. Whereas conversational technology focuses on the dialog and requires that you create a static dialog tree that limits how the employee can interact with the solution, with AI in digital employees, there is no predetermined dialog or workflow. The dialog is enabled on the fly by enabling skills. The more skills you add, the more things you can tell the digital employee to do. Additionally, conversational bots tend to be agnostic of the user. They can execute one-off stateless requests, whereas digital employees maintain a per-employee context and are able to provide a stateful, personalized experience. Lastly, digital employees are designed to be rolled out in multitudes. You could have a dozen HiRos added to an organization, each unique to the team it's working with. For example, there can be unique instances of HiRo with skills that are specific to a company's geographic region. The North America HiRo might be uniquely equipped to work with the IT systems and data specific to its geography vs. the EMEA HiRo vs. the Latin America HiRo.

RPA already automates employee tasks

Yes, and digital employees don't aim to replace RPA bots. They aim to make them accessible to employees at the time of need. Furthermore, digital employees allow employees to mix and match RPA bots, automations, and integrations to work together on the fly.

Business process and case management already orchestrate workflows

BPM and case management systems fulfill the need for end-to-end process workflow automation and provide the regulatory controls needed by enterprises for the overall business processes. Digital employees don't aim to replace these workflows but aim to fill the gaps that these workflows cannot address in a scalable way. The small tasks and the time spent accessing and working with multiple systems that consume employees daily. This is the niche digital employees aim to fill.

UP NEXT

Digital employees are the way of the future. We know it. When and how is in our hands. It's on us to shape that future responsibly, in a way that advances organizations to elevate their employees to new levels of performance and innovation.

Watson Orchestrate is IBM's point of view on the future of digital labor. Come, learn more about it and experience Watson Orchestrate firsthand.

Also, make sure you check out *The Art of Automation* podcast, especially Episode 2, in which Jerry discusses the topic of digital workers with Dr. Rania Kalaf, co-inventor of the Verdi technology mentioned in this chapter.

* * *

The next chapter covers one more fundamental topic in business automation. The intelligent document processing market is on fire and expected to grow to USD 3.7 billion in 2026, at a Compound Annual Growth Rate (CAGR) of 36.8%.[21] The major factors driving this growth include the rising need for enterprises to process large volumes of semi-structured and unstructured documents with greater accuracy and speed. Read on to explore how to put the "intelligence" into intelligent document processing.

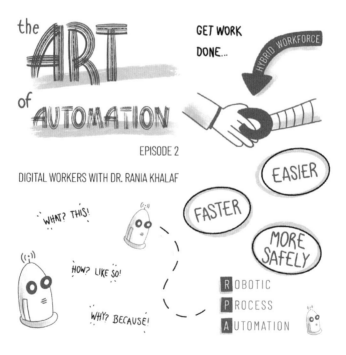

Chapter 4 – Figure 3 – Podcast Episode 2 – Automation and digital workers

Chapter 5

INTELLIGENT DOCUMENT PROCESSING

Organizations are spending more and more time manually processing documents, where we can't just blame the poor image quality of the fax machine.

Chapter Authors: Eileen Lowry and Jerry Cuomo

COVERED IN THIS CHAPTER

- The rise of AI in document processing

- Benefits of intelligent documents include better business performance

- Examples of intelligent document processing

- IBM's document processing in action

DOCUMENT PROCESSING IS RIPE FOR CHANGE

The explosion of digital content has resulted in so many variations of document formats and layouts as well as new input channels with varying quality or ability to be understood. One might be in the back seat of a ride share, trying to take a picture of a utility bill in order to apply for a time-sensitive parking permit. Or one might be exchanging emails with a patient, trying to process a healthcare claim while working from a remote home office. In 2018, Forbes stated the prior two years had generated 90% of the world's data.[22] One can only imagine how much that accelerated in 2020 between remote work, telemedicine, digital social engagements, and more.

In addition to the explosion of digital content and input channels, existing capture technology and techniques can't scale anymore. For example, document fingerprint functionality has been used to specify recognition zones (e.g., top left corner of document is defined as the "names & address zone") and positional information in order to extract the precise data needed on specific document formats or matches of similar kinds. However, with so many unique document formats coming from different channels like mobile, email, and online forms, traditional approaches to fingerprinting documents have lost their effectiveness.

The result is that organizations are spending more and more time manually processing documents, where we can't just blame the poor image quality of the fax machine. A 2019 survey conducted by Levvel Research found 57% of invoice data is entered manually and 49% of invoice approvals required two to three approvers.[23]

EMBRACING AI FOR DOCUMENT PROCESSING

While artificial intelligence (AI) is not new, it has been difficult for organizations to successfully use for processing of semi-structured and unstructured documents. Using AI has required significant data science skills and thousands of sample documents to train models. This, in turn, has resulted in long cycles to collect documents and data in order to realize business benefits.

However, advances in AI and simple tooling have been able to accelerate the use for document processing. First, deep learning algorithms have emerged, which begin to mimic the thinking of a human brain. These algorithms can identify valid contextual patterns to gain an understanding of unstructured information (like the contents of a document) and apply that learning to new documents that might be encountered in the future, which is called transfer learning. This helps reduce the document collection process and long training cycles. Second, no-code tools with simple step-by-step guides make it easy for business users to train AI models, format or convert data output, and customize business risk tolerance.

THREE MAIN ACTIVITIES OF INTELLIGENT DOCUMENT PROCESSING

While the implementation of intelligent document processing and use of AI models may differ by vendor, the core activities remain the same:

1. Document classification

2. Data extraction

3. Data output

First, *document classification* is the task by which you identify document types, such as invoices or tax forms. Using a set of sample documents, one can train an AI classification model on the different document types and the fields and values that correspond with those document types. This activity not only feeds into the next activity of data extraction but also enables transfer learning for other similar document types and facilitates better search of documents within content repositories.

Next, intelligent *data extraction* is the core activity whereby important, relevant information is pulled off the page. This consists of identifying key and value pairs like an account number or amount owed, defining what the data should look like and where it might be on the page, and training the AI models for the relevant information within each of the different document types. In this step, there may also be metadata extracted and associated with the document in order to ease search later.

Finally, *data output* consists of both enriching the data extracted and creating the final output file for use downstream. AI-based models can be used to autocorrect common misspellings, convert data into standard output formats (e.g., a telephone number) and format data to look consistent (e.g., two decimal places for dollar values). The last step is to create the output file—typically a JSON file—which can then feed a workflow or push to a content repository for use later.

DATA OUTPUT FROM INTELLIGENT DOCUMENT PROCESSING TO DRIVE PROCESS AUTOMATION

A major beneficiary of intelligent document processing is process automation, whereby structured data that has already been validated can be fed into transactions, enabling faster processing and scalable operations. For example, the manual setup of a workflow, data entry, and data validation previously may have taken hours by a human worker. An integration between intelligent document processing and workflow can eliminate these manual steps, and data output can automatically be pushed into a business process. Similarly, bad data fed into a robotic process automation (RPA) bot can result in a faulty next step, which can lead to either a bottleneck or an error in a business process. Leveraging the continuous output from intelligent document processing, an RPA bot can scale throughout an organization more easily. Finally, visualization dashboards can empower business users to uncover patterns and insights related to data extracted or bottlenecks in business processes, which can lead to more informed decision-making.

EXAMPLES OF INTELLIGENT DOCUMENT PROCESSING

There is strong evidence that there is demand for automating document processing, whereby the combination of AI and low-code tools will result in organizations improving worker productivity and driving business performance.

In fact, in working with our own IBM clients, we've uncovered a number of use cases where intelligent document processing can be applied. We'll walk through three use-case examples below and the potential benefits an organization may realize.

- *Insurance*: Account opening and servicing, personal and commercial claims

- *Government*: Social services enrollment and eligibility, pension and retirement plans, permits and licenses

- *Banking*: Account opening and servicing, mortgage/loan application

Insurance – Quote and approval process

The quote and approval process for commercial insurance is very competitive, where the first company to respond with a quote often wins the business. The challenge is that in many insurance companies, this process requires manual review, entry of application data, and reading supporting documentation, making it difficult to compete or scale. This also takes agents' focus away from advisory services, which are needed to retain and grow existing business. Intelligent document processing can automate this process using AI with deep learning to read and classify each document type and extract the appropriate data from these different formats. The extracted data can then be connected to a workflow to accelerate business processing to produce the quote and approve the application.

In this scenario, we see three specific benefits of applying intelligent document processing:

1. Processing more quotes leads to increased revenue from more business closed without the need to add staff,

2. Improved customer experience with increased processing speeds, and

3. Retention and growth of existing customer accounts.

Social services – enrollment-processing

Enrollment for dozens of local government programs, such as food assistance or subsidized housing, requires manual and tedious spreadsheet processing as IT teams do not have resources to build the required solutions. Using low-code tools and intelligent document processing, business users can build simple yet fit-for-purpose processing applications and train the system to recognize key fields from enrollment forms. In addition, easy-to-configure validators can ensure date fields and currency fields are accurately recognized, and simple custom validators can also be created to handle unique fields like a social security number.

Like the prior scenario, there are three benefits of applying intelligent document processing:

1. Increased program enrollment due to faster turnaround times,

2. Cost-effective rollout of custom automation solutions with appropriate role-based viewing of personally identifiable information, and

3. Built by business users with little to no involvement from IT.

Banking – Personalized account servicing

Banks can have over twenty different account servicing forms available for download from the Banks website. Account holders use these forms to make changes to accounts or close accounts. Today, this can require a sizable team of agents to read these forms, verify the data, and then enter the data into an account management system. However, with low-code tools and intelligent document processing, the bank can rapidly build solutions to process each account servicing form and use intelligent document processing to train the system on each form in order to recognize not only common fields like customer address and account number but also unique fields to each form.

By combining with RPA, banks can also take the extracted data and automate the changes into their own backend systems. Additionally, leveraging intelligent document classification, account-closing forms can quickly be flagged and agents alerted to clients that may be potential flight risks.

Similar in nature to the prior scenarios, applying intelligent document processing provides the following benefits:

1. Improved customer experience with faster response times,

2. Better customer retention with intelligent flight risk identification, and

3. Reduced retail banking costs on a per-account basis.

IBM AND INTELLIGENT DOCUMENT PROCESSING

IBM's approach to intelligent document processing surfaces in our IBM Cloud Pak® for Business Automation. A cloud-native solution, Automation Document Processing is a set of AI-powered services that automatically read and correct data from documents. A document processing designer provides an easy-to-use no-code interface for training models on document classification, data extraction, and data enrichments.

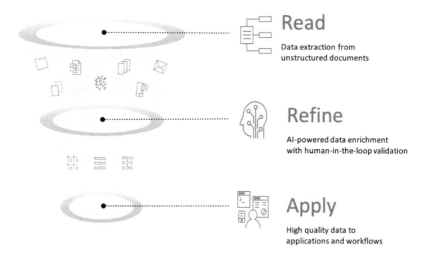

Read

Data extraction from unstructured documents

Refine

AI-powered data enrichment with human-in-the-loop validation

Apply

High quality data to applications and workflows

Chapter 5 – Figure 1: Automating document processing

In addition, IBM provides document processing application templates that can be used for processing either single-page documents or batches of documents. Toolkits in the application designer can also be used to customize the end-user application to look and feel like other applications within an organization. Finally, IBM provides simple deployment tools and an out-of-the-box integration with its content services capabilities, IBM FileNet Content Manager, for both storing the document(s) and data output file.

INTELLIGENT DOCUMENT PROCESSING IN ACTION

This section provides a quick demonstration of intelligent document processing in action. The use case is a *utility bill payment application.*[24] In this example, banking customers simply forward their utility bills to their bank via email. All the relevant information contained in the utility bills, such as payment amount and due date, are extracted and validated using IBM's automation document processing. The extracted information is then fed into the bank's back-office system for processing automatic payments on behalf of the customers.

In this example, we show how Bill, a business analyst at ABC Bank, uses document design and build tools to set up a document model and application UI using an out-of-the-box template. Then Buffy, a data clerk at ABC Bank, reviews "flagged" documents that require further attention to continually train the system to perfection.

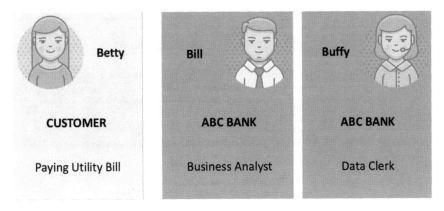

Chapter 5 – Figure 2 – ABC Bank personas

The following figure shows the document designer tool. Bill is using this tool to create a document model for a "utility bill" by defining fields like "customer number," "customer name,", and "amount due." Bill did not have to manually define all these fields from scratch. These fields were automatically extracted using a "smart template," which was trained by examining hundreds of bills. As Bill refines the model, the machine learning model behind the utility bill document gets smarter.

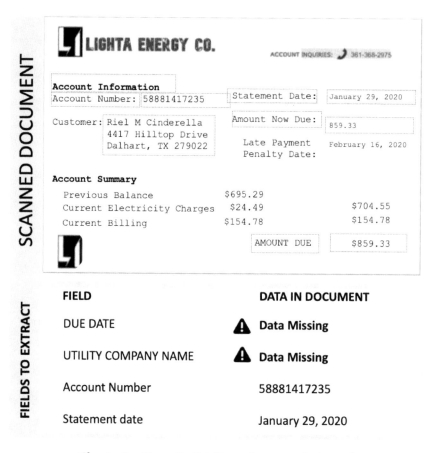

Chapter 5 – Figure 3 – Intelligent document design tool

Bill needs to make a refinement to the field "due date," which is flagged as "data missing!" This may be because the model has yet to be trained to recognize the multiple ways dates might be represented in a bill (e.g., February 16, 2020 vs. 02/16/2020). To correct this, all Bill does is teach the system this variation of date by defining a new syntax in the training tool. Bill makes other corrections, including additional training on how to recognize variation in "customer name," which sometimes incorrectly seems to pick up the "customer address" as part of the "customer name." Bill uses a "split into composite fields" option to cleanly separate name and address.

Bill does not have to do these manual corrections for every single utility bill. He just needs to do this for a few samples for the deep learning algorithm to be updated. Once done, the system is smarter and will automatically apply these improvements to other utility bills as they are encountered.

The next figure shows how validation logic can be defined or customized. Default validators are provided, such as low confidence, data type mismatch, and required value. If a document value does not meet the criteria defined by the validator, it will be flagged for a human to further investigate. In the case of ABC Bank, that's Buffy's job as we will see shortly.

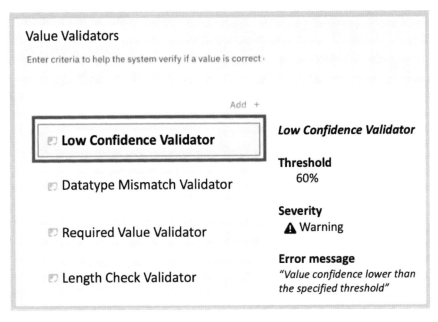

Chapter 5 – Figure 4 – Validation logic

The figure below shows a utility bill that has a validation exception flagged. Buffy can see that the "total amount due" field was flagged because its value is too large. Bill had specifically trained the system at design time with validation logic to make sure the total amount due for the utility bill isn't an abnormally high or low amount. Buffy can investigate and collaborate with Bill to further train the utility bill document model to make it even smarter.

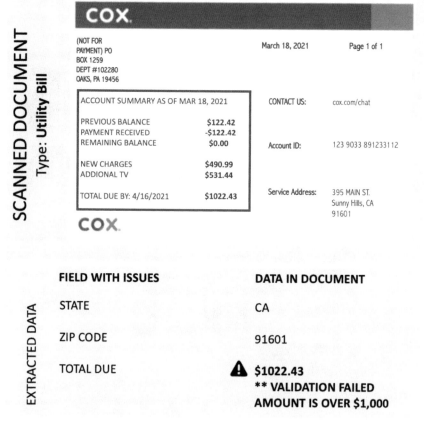

Chapter 5 – Figure 5 – Utility bill data checking application

With this demonstration, you can easily imagine how ABC Bank can fully automate the processing of thousands of similar utility bills without human intervention, thereby reducing cost and errors associated. But more importantly, Betty, the customer of ABC Bank, is delighted because she can now pay her bills on time and with ease. And that's a win-win!

THE FUTURE OF INTELLIGENT DOCUMENT PROCESSING

While this chapter gave an overview of how document processing has been ripe for change and where AI is playing a major part in advancing document processing, there is more innovation to come in this space. There are two key areas, in particular, to keep an eye out on. First, as the formats and structures of semi-structured and unstructured documents continue to explode, AI models will need to keep up. From reading highly complex table structures to processing government-issued IDs with holograms or watermarks, AI models will be challenged to remain accurate.

Second, while this space has been coined intelligent *document* processing, video and audio file types are on the rise. It is only a matter of time before these file types are in the critical path for processing of insurance claims or filing of police incident reports.

Stick around for the ride; it is sure to be exciting.

UP NEXT

Make sure you check out *The Art of Automation* podcast, especially Episode 7, in which Jerry and Eileen discuss why document processing is so manually intensive and what makes it so difficult to automate. Eileen describes how AI/deep learning can mimic the human brain to understand the unstructured information found in an enterprise's wide variety of documents. She shares an example from the insurance industry that shows the enormous business value of intelligent document automation.

* * *

If *business automation* is the right side of an autonomous enterprise's brain, then *IT automation* is the left side. The prior chapters explored the most fundamental topics related to business automation. In the next set of chapters, experts in IT automation will explore the "left side," starting with observability. As you read these chapters, it is important to appreciate that to truly become an autonomous enterprise, the left and right sides of the brain must work together. The linking of business and IT automation is covered in the Business Ops (BizOps) section of the next chapter. Read on....

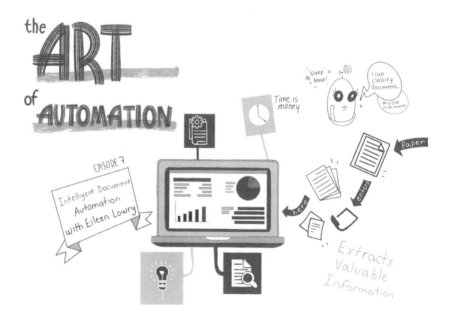

Chapter 5 – Figure 6 – Podcast Episode 7 – Automation and intelligent document processing

Chapter 6
OBSERVABILITY

What's needed is higher quality telemetry—and a lot more of it—that can be used to create a high-fidelity, context-rich, fully correlated record of every application user request or transaction.

Chapter Author: Jerry Cuomo

COVERED IN THIS CHAPTER

- What is observability?

- Benefits of observability

- The future of observability

- The ARM of observability

TURNING THE LIGHTS ON?

There's a saying… "You can't fix what you can't see." Seems logical, right? Well, that saying also translates well to automation, as in "you can't automate what you can't see." Strictly speaking, you probably could "automate blindly." But then again, what are the odds that you would fix the issues that matter most to your business? "Turning the lights on" in the context of automation means having *visibility* to the data that, when put in the right context, will lead your business to the most rewarding use of automation technology.

Observability provides deep visibility into modern distributed applications for faster automated problem identification and resolution. In Episode 9 of *The Art of Automation* podcast, Mirko Novakovic, CEO of Instana, defines observability as an evolution of application performance monitoring (APM) that is "the art of understanding what is happening inside an application, from the outside," resulting in observability being the "data source for automation."

With this simple definition, we can start to appreciate the key elements of observability and what it can deliver to a business. In this chapter, we will dive deeper into observability and its relationship to automation. Basic definitions are provided, including an overview of whom observability benefits and how it works. The chapter also provides a detailed example of observability in action using IBM Instana and concludes with a quick look into the near future of observability.

WHAT IS OBSERVABILITY?

In general, observability is the extent to which you can understand the internal state or condition of a complex system based only on knowledge of its external outputs. The more observable a system, the more quickly and accurately you can navigate from an identified performance problem to its root cause, without additional testing or coding.

The term "observability" comes from control theory—an area of engineering concerned with automating a dynamic system (e.g., the flow of water through a pipe or the speed of an automobile over inclines and declines) based on feedback from the system.[25]

In cloud computing, observability also refers to software tools and practices for aggregating, correlating, and analyzing a steady stream of performance data from a distributed application and the hardware it runs on. This allows you to monitor, troubleshoot, and debug the application to meet customer experience expectations, SLAs, and other business requirements more effectively.

A relatively new IT topic, observability is often mischaracterized as an overhyped buzzword or a "rebranding" of system monitoring, or more specifically, application performance monitoring (APM). In fact, observability is a natural evolution of APM data-collection methods that better addresses the increasingly rapid, distributed, and dynamic nature of cloud-native application deployments. Observability doesn't replace monitoring—it enables better monitoring and better APM.

WHY DO WE NEED OBSERVABILITY?

For the past twenty years or so, IT teams have relied primarily on APM to monitor and troubleshoot applications. APM periodically samples and aggregates application and system data—called *telemetry*—that's known to be related to application performance issues. It analyzes the telemetry relative to KPIs and assembles the results in a dashboard to alert operations and support teams to abnormal conditions that should be addressed to resolve or prevent issues.

APM is effective enough for monitoring and troubleshooting monolithic applications or traditional distributed applications, where new code is released periodically and workflows and dependencies between application components, servers, and related resources are well known or easy to trace.

But today, organizations are rapidly adopting modern development practices—agile development, continuous integration and continuous deployment (CI/CD), DevOps, multiple programming languages—and cloud-native technologies like microservices, Docker containers, Kubernetes, and serverless functions. As a result, they're bringing more services to market faster than ever. But in the process, they're deploying new application components so often, in so many places, in so many different languages, and for such widely varying periods of time (for seconds or fractions of a second, in the case of serverless functions) that APM's once-a-minute data sampling can't keep pace.

What's needed is higher quality telemetry—and a lot more of it—that can be used to create a high-fidelity, context-rich, fully correlated record of every application user request or transaction. Enter *observability*.

WHO BENEFITS FROM OBSERVABILITY?

Many roles across a modern enterprise benefit from observability. DevOps and site reliability engineering (SREs) teams are likely the most immediate benefactors. However, when you consider that most IT roles today are tied to the success of business performance and customer satisfaction, few things impact these factors more than application performance—when an application is not performing, the business is not successful, or customers are not happy. Hence, any major IT role benefits from the real-time insight gained from observability software.

Later in this chapter, we introduce the notion of BizOps. With BizOps' features, observability solutions are evolving to benefit business leaders in a way that aligns technology efforts and investments with strategic business objectives that simply deliver better results, faster, and this is what observability is all about.

In this chapter, the role of SRE is singled out because of the versatility of the role and how an observability platform acts as a digital assistant—offloading the tedious tasks of instrumenting code and collecting data while analyzing logs, metrics, and traces. In general, an SRE team is responsible for application availability, latency, performance, efficiency, change management, monitoring, emergency response, and capacity planning.[26] They split their time between operations/on-call duties and developing systems and software that help increase site reliability and performance. The automation provided by observability software helps them to focus on higher value tasks related to the well-being of the enterprise.

HOW DOES OBSERVABILITY WORK?

Observability platforms continuously discover and collect performance telemetry by integrating with existing instrumentation built into application and infrastructure components and by providing tools to add instrumentation to these components.

Observability focuses on four main telemetry types:

- *Logs*: Logs are granular, timestamped, complete, and immutable records of application events. Among other things, logs can be used to create a high-fidelity, millisecond-by-millisecond record of every event (complete with surrounding context) that developers can "play back" for troubleshooting and debugging purposes.

- *Metrics*: Metrics (sometimes called time series metrics) are fundamental measures of application and system health over a given period of time, such as how much memory or CPU capacity an application uses over a five-minute span or how much latency an application experiences during a spike in usage.

- *Traces*: Traces record the end-to-end "journey" of every user request, from the UI or mobile app through the entire distributed architecture and back to the user.

- *Dependencies*: Dependencies (also called dependency maps) reveal how each application component is dependent on other components, applications, and IT resources.

After gathering this telemetry, the observability platform correlates it in real time to provide SRE teams with contextual information—the what, where, and why of any event that could indicate, cause, or be used to address an application performance issue.

Many observability platforms automatically discover new sources of telemetry that might emerge within the system (such as a new API call to another software application). As new sources emerge it becomes increasingly difficult to correlate data and events across these sources. Therefore, many observability platforms include AIOps (artificial intelligence for operations) to provide capabilities that sift the signals (indications of real problems) from noise (data unrelated to issues).

BENEFITS OF OBSERVABILITY

The overarching benefit of observability is that with all other things being equal, a more observable system is easier to understand (in general and in great detail), easier to monitor, easier and safer to update with new code, and easier to repair than a less observable system. More specifically, observability directly supports the Agile/DevOps/SRE goals of delivering higher quality software faster by enabling an organization to do the following:

- *Discover and address "unknown unknowns" (issues you don't know exist)*: A chief limitation of monitoring tools is that they only watch for "known unknowns"—exceptional conditions you already know to watch for. Observability discovers conditions you might never know or think to look for, then tracks their relationship to specific performance issues and provides the context for identifying root causes to speed resolution.

- *Catch and resolve issues early in development*: Observability bakes monitoring into the early phases of software development process. DevOps teams can identify and fix issues in new code before they impact the customer experience or SLAs.

- *Scale observability automatically*: With observability, an SRE, for example, can specify instrumentation and data aggregation as part of a Kubernetes cluster configuration and start gathering telemetry from the moment it spins up, until it spins down.

- *Enable automated remediation and self-healing application infrastructure*: Combine observability with AIOps machine learning and automation capabilities to predict issues based on system outputs and resolve them without management intervention.

OBSERVABILITY IN ACTION

With Instana, IBM offers state-of-the-art AI-powered automation capabilities to manage the complexity of modern applications that span hybrid cloud landscapes—especially as the demand for better customer experiences and more applications impacts business and IT operations.

Any moves towards business-wide and IT-wide automation should start with small, measurably successful, projects, which you can then scale and optimize for other processes and in other parts of your organization. By making every IT services process more intelligent, teams are freed up to focus on the most important IT issues and accelerate innovation.

This section provides a deeper dive into observability through the lens of an example using IBM Instana as the observability tool of choice.[27]

One of the core tenets of effective APM is to maximize the amount of visibility with the least amount of effort. This is where Instana really shines, recognizing over a hundred discoverable IT components running in an IT environment without having to be configured or programed.

Auto-discovery of IT components

The following illustration is of an infrastructure-map dashboard, where each of the blocks or cubes represents a host or node that is being monitored. The figure shows one particular cube being hovered over, displaying the components that were auto-discovered in the selected Kubernetes cluster. Instana sensors auto-discovered the entire Docker runtime, and then within the containers, it discovered application and IT components, including Spring Boot apps, Rabbit MQ messaging brokers, Java Virtual Machines (JVMs), and Elasticsearch. Again, observability is about maximizing visibility with minimal effort, so discovering these IT components automatically allows users to increase the frequency of deployments without friction or drag on their DevOps pipelines.

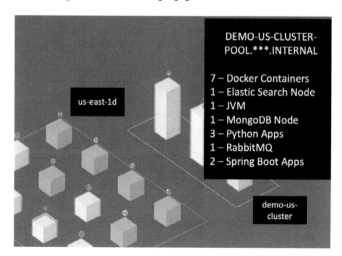

Chapter 6 – Figure 1 – Instana's Infrastructure-Map Dashboard

Observing components to IT environment dependencies

The next illustration shows how a better understanding can be gained about the relationship of an IT component and their impact on the IT environment in which it is running. This is accomplished by providing visibility into key aspects of the operating system and related infrastructure metrics being collected for every host, including CPU usage, memory usage, open files, and network IO information. This base set of infrastructure metrics is important and provides the vital signs that are driving underlying anomaly detection routines, which will be discussed later in this chapter.

In the lower-left side of the illustration, the Spring Boot application is highlighted. One could drill into its dashboard to show additional metrics like Requests and HTTP sessions related to this Spring Boot app. Furthermore, the lineage and relationship are mapped between the Spring Boot app, the JVM it's running on, the Java process container and the Kubernetes pod and corresponding infrastructure node. This sets the foundation for establishing context by which activities can be better correlated and pinpointed, as we will see in the next section.

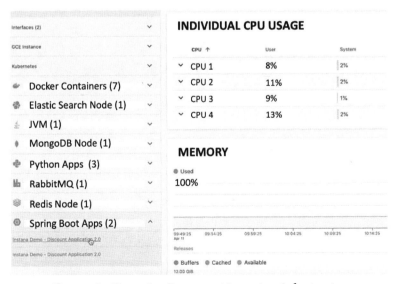

Chapter 6 – Figure 2 – Component impact on infrastructure

Observing components to application dependencies

Most users have little trouble conceptualizing an application—what it does and how to observe it—but the truth is that pinning down the exact makeup of an application is tricky because applications are often a loosely coupled collection of IT components. A key role of an observability tool is to help a user piece together the relevant IT components that constitute an application as observed by an end user. Hence, if a user observes that their shopping app is down, then the SRE must be able to see "the forest from the trees." In other words, the SRE must be able to understand what IT components (trees) make up an application (forest), as well as the dependencies that these components have on each other.

The next illustration shows the IT components that have been discovered by Instana. In this case, we see a set of web services, databases, caching engines, and async message brokers. Again, discovery automatically occurs for most of today's application runtime languages. Rather than having developers instrument their code or provide hints in the form of code annotations, sensors automate the identification of component names, along with the type of runtime (e.g., database) and brand of runtime technologies that are affiliated (e.g., MySQL). The illustration below shows how sensors recognize a complete list of IT components, which is the first step to piecing together an application jigsaw puzzle.

For every IT component that is discovered, Instana focuses on collecting telemetry data across three "golden signals," which are three KPIs including the *invocation rate* (calls), *latency*, and *erroneous call rate*. These KPIs are critical because they ultimately feed the anomaly-detection algorithm that is the intelligence behind this observability tool. Therefore, if there is a spike in error rates or an increase in latency, the SRE can be alerted to that fact and take appropriate action.

Name	Types	Technology	Application	Calls ↓		Latency ↓
nginx-web	HTTP	ⓞ Nginx	4	60,846		21ms
catalogue-demo	HTTP	ⓞ Spring Boot	5	40,215		14ms
catalogue	Database	✦ MongoDB	6	40,213		< 1ms
redis:6379	Database	⬙ Redis	3	19,824		1ms
discount	HTTP	ⓞ Spring Boot	6	15,383		24ms
ratings	Database	⬚ MySQL	6	15,282		1ms
MySQL@3306 on demo-	Database	⬚ MySQL	7	14,031		21ms
cart	HTTP	ⓢ Node.JS	6	11,097		17ms
shipping	HTTP	⬚ JVM	6	8,778		48ms

Chapter 6 – Figure 3 – View of IT components and relationship to application

Modern observability platforms start to connect the dots and correlate the relationships between these IT components, such that the forest emerges from the trees. Besides collecting the golden KPIs, a distributed trace attempts to show exactly how the dots connect. A distributed trace is a specific diagnostic view into the individual request calls between these IT components. Data collection is the foundation for distributed tracing. Every application call is collected while correlation algorithms are applied to start mapping relationships across all participating IT components. So, if there is an application issue in production, a trace is always available for reference to help SREs understand what is going on for that issue.

Once IT components are discovered and distributed tracing has begun, Instana will start to group the components into applications. The following is an illustration of an application-level dashboard of the sample Robot Shop application. This view aggregates the golden KPIs across all IT components grouped into Robot Shop.

Tight correlation between the IT components and application infrastructure provides immediate visibility of processing time and latency, providing the foundation a variety of insights into the performance and health of the application.

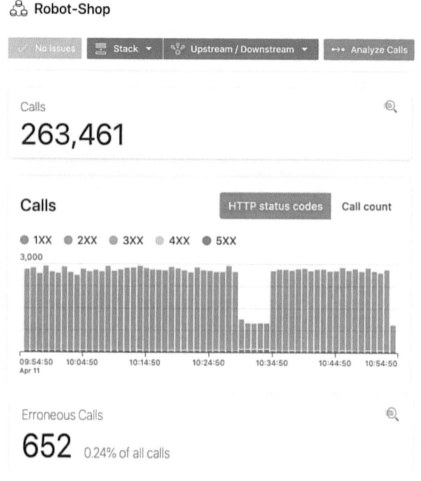

Chapter 6 – Figure 4 – Robot Shop application's affiliated IT components

Now that detailed application dependencies are clearly understood with telemetry data actively collected and traced, a dependency map—as shown in the next illustration—is a powerful visualization of the application and acts as a sanity check of your application architecture. However, unlike a Visio drawing of your application, this map is live and allows the heart of your application (i.e., the calls between IT components) be observed in real time:

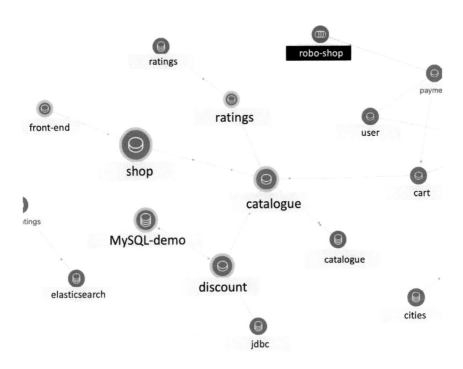

Chapter 6 – Figure 5 – Robot Shop's active component dependency map

By collecting and correlating all log messages that are occurring in the Robot Shop app, the observability tool has taken a giant step towards detecting anomalies. Specifically, by applying analytics to associated log data, it is possible to identify all the error messages for Robot Shop. If the SRE was just working with log files, she might be lucky and get a decent description of a particular error. However, what she won't get out of standard log files is the context in which that error occurred. She can focus on getting her triage and remediation started quickly because Instana has connected the error to calls occurring before and after the error. The following illustration shows an example of a call graph leading up to a particular application error in Robot Shop.

Chapter 6 – Figure 6 – Call graph pointing to anomaly

In the sample illustration above, it appears that there is an error to do with database connections. The SRE can drill down into that error and list of all the traces where this particular error occurred. This distributed trace provides diagnostic visibility of the services call graph, including the timings involved and where these errors were occurring. Additional detailed information on the error appears in the right-hand part of the screen that attempts to pinpoint the root cause of the error. In this case, the last call before the error was to a database service that is requesting a connection to a MySQL database over a JDBC connection. With this level of diagnostic visibility, the SRE gets immediate context in real time, providing a much quicker resolution of production incidents.

The next section shows how Instana can further automate the detection and remediation of anomalies.

AUTOMATED ANOMALY ALERTS DELIVERED WHERE YOU WORK

The example above illustrates how observability maximizes visibility with the least amount of effort. What was once manually done by human developers and IT operators is now automated. We can now sense and discover using intelligent software to give us full visibility into the vital signs that keep the Robot Shop application delivering value to our business.

However, the value delivered to an SRE can be further multiplied with additional automation that provides insight on incident root cause. From here, observability starts to trigger automated actions that ultimately change SRE from a reactionary practice to proactive practice. Such automations can allow anomalies to be detected and remediated before they cause damage to your business.

In this section, we further show how Instana's anomaly detection algorithms detect situations and instantly alert us where we work. Instana has a library of built-in events that represent the heuristics and behavior of well-known anomalies that commonly occur for specific service types. SREs do not need to sit in front of an observability dashboard 24×7 waiting for metrics to "turn red." For this, Instana supports alert channels, including email, Slack, and Microsoft Teams.

For example, as Robot Shop is being used in production, Instana is observing telemetry data and applying event heuristics to look for anomalies, like a sudden increase in database connection errors when customers access the Robot Shop catalog. When this occurs, the SRE on call will receive, say, a Slack message notifying them of the health issue in Robot Shop, with a summary of the issue and hyperlink for more information. When the SRE clicks on that hyperlink, she is taken to an Event Viewer, like the one showed in the illustration below.

The Event Viewer gives all the information needed to ultimately determine and remediate the root cause. This event involves a spike in errors emitted from the Robot Shop catalog-demo service. Instana suggests that the MySQL database that is related to the catalog-demo service has abnormally terminated because it was not able to acquire more memory. Furthermore, a spike in user requests to view the catalog seems be the reason why the database needed more memory. In this case, the SRE determines this to be the root cause and increases the memory to the MySQL database service.

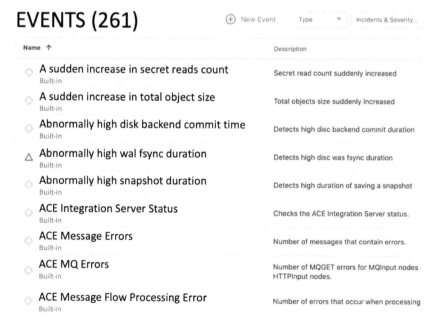

Chapter 6 – Figure 7 – Event Viewer with classified anomalies

In this example, Instana acted as an intelligent digital assistant to the SRE. It automated many manual troubleshooting tasks and provided insight into the issue with context to the application in relationship to its sub-components. In short, Instana served up the incident to the SRE on a silver platter, such that she was able to spend less time investigating and more time remediating. An outage that may have taken hours to troubleshoot only took minutes, minimizing the impact on business and customer satisfaction.

THE FUTURE OF OBSERVABILITY

In this chapter, we illustrated how an observable system is easier to understand, easier to monitor, easier and safer to update with new code, and easier to repair than a less observable system. Automating the mundane tasks of sensing, collecting, and correlating IT telemetry data frees an SRE to keep business applications performing, which is a key contributor to customer satisfaction.

We saw how observability is a natural evolution of APM that better addresses the increasingly rapid, distributed, and dynamic nature of cloud-native application deployments. We also saw examples of how observability places incidents into a context that pinpoints the application and the specific IT sub-components that are related to the incident and how using event heuristics can match the incident to a well-known root cause and remediation. Such capabilities radically reduce the time to recover from hours to minutes, which has immense value to an enterprise.

This section concludes this chapter with a look at three compelling complements to observability that are examples of what's in store in the near future.

The ARM of observability

As observability solutions evolve, they are growing more versed at automating actions based on what is observed. An example of this is IBM's

Turbonomic, which is an Application Resource Management (ARM) tool. The combination of APM and ARM starts to illustrate how observability can more aggressively automate actions based on insights. This topic is covered in detail in Episode 16 of *The Art of Automation* podcast: The ARM of Automation with Ben Nye.

ARM helps improve application performance by ensuring applications get the resources they need to perform by automating IT resource provisioning to prevent resource congestion. With enterprise applications running on autopilot, SRE teams can further shift their energy to innovation and reclaim time to drive better customer experiences.

AIOps

Observability and AIOps are intimately linked and will continue to form an even stronger bond in the near future. Observability platforms are often viewed as a primary data source (logs, metrics, traces, dependency graphs, etc.) for the training and machine learning algorithms that can transform IT from reacting to incidents to proactively responding to predictions of outages that have yet to occur, thereby avoiding the incident altogether, including the cost and time associated. It can be viewed that observability is the "nervous system" that delivers signals to the AIOps "brain." More on this topic is included in the AIOps chapter to follow, written by Rama Akkiraju.

The next topic describes how observability can be extended to allow SREs to better collaborate with their business counterparts to align IT performance with business performance.

Business Operations (BizOps)

Today's observability solutions tend to be focused on providing visibility to IT systems. However, almost every IT system exists to power a business system or process. That said, we see the future of observability moving up the value chain to include business observability. Think of this as APM meets BPM (business process management). Business Operations (BizOps) is a field that extends SRE to provide oversight of the well-being of business processes. With business observability, a business leader can instantly understand the impact of an IT incident on their business, measured in terms of cost, time, net promoter score, etc.

Business observability introduces business telemetry via sensors that emit business events from BPM, business rules, content management, and process mining software. The mix of business and IT telemetry, including the correlation and dependency mapping that is indigenous to observability, allows business incidents to be detected and remediated by pinpointing the root causes within the IT systems that is powering business processes. BizOps solutions often include technology to create dashboards that provide a real-time visualization of business processes, KPIs, and their relationship and impact on IT metrics.

For example, say an insurance business process related to medical claims is missing its SLA, which is to process each claim within twenty-four hours. Business observability would be able to connect the dots between tasks in the business process, say a "claims approval" task, and their underlying IT service that is powering that task. Perhaps the root cause of this business incident is a message-queue (i.e., and IT service) running out-of-disk-space. With business observability, the IT and business team (say the customer support group) will be aligned because they will simultaneously discover (via a dashboard or alert delivered via Slack) the incident, allowing customer support to proactively reach out to the customer of the medical claim and process immediately, while the SRE team remediates by adding more storage. Result? A resilient business leading to happy customers and outstanding business performance.

The future of observability brings IT and business telemetry together and uses AI to sense and respond to incidents before they happen, while capitalizing on opportunities ahead of your competition. By establishing a new level of context that spans business and IT silos of the past, observability is evolving to enabling unparalleled levels of cross-team visibility and collaboration. Aligning technology efforts and investments with strategic business objectives—simply deliver better results, faster—and this is what observability is all about.

UP NEXT

Make sure you check out *The Art of Automation* podcast, especially Episode 9, which is directly related to this topic. In this episode, Jerry is joined by CEO of Instana, Mirko Novakovic. They discuss how the increasing complexity of AI and cloud environments can lead to the creation of an enterprise black box, where no one knows exactly how an application works or how AI models make decisions. Mirko describes how APM and observability play a critical role in tackling this problem for an enterprise by helping them "turn on the lights." He also introduces "Stan" (the robot) to discuss the future of automation in observability.

* * *

Observability sets the table for the next chapter on the popular topic of AIOps. Similar to other topics covered in this book, observability and AIOps are closely related. One might argue, there is a natural "crawl, walk, and run" progression that can provide an orderly sequence to the usage of APM, ARM, and AIOps. So, now that we've crawled and walked, we will be running ahead in the next chapter.

Chapter 6 – Figure 8 – Podcast Episode 9 – Automation and observability

Chapter 7

AIOps

We envision fully instrumented, observable, self-aware, automated, and autonomic IT operations environments in the future. AI can help us realize the proverb "Prevention is better than cure."

Chapter Author: Rama Akkiraju

COVERED IN THIS CHAPTER

- The vision of an autonomic IT system

- Holistic approach to IT operations management

- Transformation of ITOps to AIOps

- The future of AIOps "shifts left"

PROGRESSING TOWARDS AN AUTONOMIC IT SYSTEM

The vision of self-aware, self-healing, and self-managing IT systems has remained elusive until recently. Recent advancements in cloud computing, natural language processing (NLP), machine learning (ML), and artificial intelligence (AI) in general are all making it possible to realize this vision now. AI can optimize IT operations management processes by increasing application availability, predicting and detecting problems early, reducing the time to resolve problems, proactively avoiding problems, and optimizing the resources and cost of running business applications on hybrid clouds.

In this chapter, we detail the opportunity for AI in IT operations management and the techniques that we are developing at IBM as part of the IBM Cloud Pak® for Watson AIOps product. We will describe how semi-structured application and infrastructure logs are analyzed to predict anomalies early; how entities are extracted and linked from logs, alerts, and events to reduce alert noise for IT operations admins; how NLP is put to use on unstructured content in prior incident tickets to extract next-best-action recommendations to resolve problems; and how deployment change request descriptions are analyzed in combination with past incident root-cause information to predict risks of deployment changes to prevent issues from happening in the first place.

IT OPERATIONS MANAGEMENT

Information Technology (IT) Operations management[28] is a vexing problem for most companies that rely on IT systems for mission-critical business applications. Despite the best intentions of engineers, good designs, and solid development practices, software and hardware systems deployed in companies in service of critical business applications are susceptible to outages, resulting in millions of dollars in labor, revenue loss, and customer-satisfaction issues each year. According to a recent *Forbes* article, every year, IT downtime costs an estimated $26.5 billion in lost revenue based on a survey of two hundred companies across North America and Europe.[29]

The best of the analytical tools falls short of detecting incidents early, predicting when incidents may occur, offering timely and relevant guidance on how to resolve incidents quickly and efficiently and helping avoid them from recurring. This can be attributed to the complexity of the problem at hand.

Data volumes continue to grow rapidly as companies move to modular microservices-based architectures, further compounding the problem. Gartner estimates that the data volumes generated by IT infrastructure alone are increasing twofold to threefold every year. Furthermore, the heterogeneous nature of environments—where companies' IT applications can run on a mix of traditional bare metal, virtual machines, and public or private clouds operated by different parties—adds to the complexity and scale that IT operations management solutions must deal with.

To add to this, IT applications, the infrastructure that they run on, and the networking systems that support that infrastructure all produce large amounts of structured and unstructured data in the form of logs, traces, and metrics. The volume and the variety of data generated in real time pose significant challenges for analytical tools in processing them for detecting genuine anomalies, correlating disparate signals from multiple sources, and raising only those alerts that need IT operations management teams' attention. Having the best-of-breed IT operations management tools is necessary, but not sufficient, for effective problem resolution. Such complex and dynamic environments demand a new approach to IT operations management that is smart, intelligent, real-time, adaptive, customizable, and scalable.

AI can help solve these problems. AI can help IT operations administrators, also known as Site Reliability Engineers (SREs), in detecting issues early, predicting them before they occur, reducing event and alert noise, locating the specific application or infrastructure component that is the source of the issue, determining the scope of incident impact, and recommending relevant and timely actions. All these analytics help reduce the mean times to detect (MTTD), mean times to identify/isolate (MTTI), and mean times to resolve (MTTR) an incident. This, in turn, saves millions of dollars by preventing direct costs (e.g., lost revenue, penalties, opportunity costs, etc.) and indirect costs (e.g., customer dissatisfaction, lost customers, lost references, etc.).

The rest of the chapter is organized as follows. First, we describe our holistic approach to IT operations management, elaborating on the broader opportunity for AI to optimize various problems in the domain. We then focus specifically on what we at IBM are doing to achieve the vision of self-healing, self-managing, and self-monitoring IT systems. Finally, we conclude by reiterating our vision and the opportunity at hand.

A HOLISTIC APPROACH TO IT OPERATIONS MANAGEMENT

AI enables us to take a holistic approach to IT operations and service management. This is referred to as AIOps. We elaborate our vision for applying AI to optimizing IT operations management in the following figure.

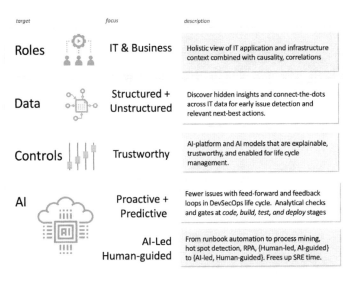

Chapter 7 – Figure 1: Our vision for AIOps

From structured to structured, semi-structured, and unstructured

Traditionally, the primary approach to addressing IT operations issues has been by monitoring the metrics, which are structured data. However, unstructured data like logs and prior incident ticket data (which is semi-structured and unstructured) can help detect issues early and resolve problems based on prior resolutions. The rise of AI—powered by the advancements in hardware architectures, cloud computing, natural language processing (via language models like BERT[30] and Fasttext[31]), machine learning (via deep learning [DL] algorithms and frameworks like Tensorflow and Pytorch), and deep neural network architecture optimization frameworks (like Katib)—has opened up new opportunities for process unstructured data.

We can now pre-train features in multiple languages using language models.[32] We can extract noun–verb phrases from prior ticket incidents to identify resolutions.[33] We can apply semantic parsing techniques to extract key terms and phrases to derive runbooks.[34] Using these latest NLP techniques, we can now untap the potential from logs and tickets in the IT operations domain to detect signals and problem resolutions.

From siloed signals to integrated context

By extracting mentions of the problem components (e.g., entities like application names, server names, pod ids, node ids, etc.) from various structured and unstructured data, we can connect the dots across IT data and create a holistic problem context. When combined with topology and causality reasoning, this correlation of data across various signals can help us create a full picture of the context around a problem, thereby facilitating better problem resolution.

From reactive to predictive and proactive

"Prevention is better than cure," according to an old proverbial saying. In our view, IT operations management and service management must include not only monitoring of business applications and optimizing incident management and problem resolution but also designing IT systems, applications and services, building them, testing them, and deploying them with the highest quality possible so as to avoid problems in the first place. In essence, design to operate better from the get-go.

We envision various stages of IT application development processes and tools for coding, building, testing, deploying, and monitoring to be equipped with AI-infused smarts to guide developers, testers, deployment engineers, and IT operations engineers (also referred to as SREs) to write secure, stable, and scalable software to start with.

If problems were to still trickle through—which they will, as it may not be possible to catch every problem during code, build, test, and deploy—we envision catching them at the end of each stage via risk-prediction models to prevent poor-quality artifacts that do not meet the preset quality criteria from getting promoted to the next stage. For example, smart checks and gates block code with risky security vulnerabilities from getting to the deployment phase, stop under-tested code modules from getting into deployment phases, prevent risky deployments from getting pushed to production, and so on.

We envision our AIOps solution correlating past incidents with root causes that could be traced to security vulnerabilities, poor code test coverage, and under-tested deployment changes. This information, when fed back, serves as critical input to reinforcing the checks and gates in the earlier stages of the Development-Security-Operations (DevSecOps) lifecycle, as shown in the following figure.

Feedback loops for predictions & process
improvements at every phase

E.g., flag or block risky application or configuration changes

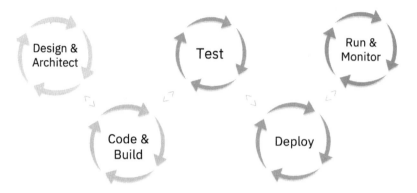

Feedforward loops better inform
subsequent phase analytics

E.g., knowledge of changes and their risk assessment, accelerates incident
management of downstream issues.

Chapter 7 – Figure 2: Shifting left in the DevSecOps lifecycle

In the following table, we describe some of the analytics that AI can drive
in incident management use cases:

Reactive Analytics	Predictive Analytics	Proactive Analytics
Log and metric anomaly detection	Log and metric anomaly prediction	Deployment change risk prediction
Event grouping	Incident severity prediction	Code vulnerability prediction
Fault localization	Incident classification	Test coverage analysis
Blast radius	Incident impact prediction	Root cause analysis
Incident summarization and query formulation	Root cause prediction	Runbook derivation
Incident similarity and next-best-action recommendation	Incident resolution complexity and time prediction	Process mining, process analysis

Chapter 7 – Table 1: AI analytical opportunities in IT operations management

FIRST-GENERATION AI MODEL MANAGEMENT TO ADVANCED AI MODEL MANAGEMENT

Prediction models built using machine learning are bound to make prediction mistakes. They need to learn on the job and keep improving. It's one thing to build AI models and to deploy them in production, but it's a whole different thing to build them in such a way that the models are learning continuously and improving from fresh, fair, balanced, and unbiased data taking user feedback as it comes. For this, AI models need to conduct disciplined error analysis from each iteration. Having an AI platform that supports the management of the lifecycle of AI models is critical so that AI models are fresh and relevant.

Such a platform should support both the data scientists that build the initial models and AIOps product and IT operations tool administrators that have to maintain these AIOps tools in production. These IT operations tool administrators are not data scientists; so, care must be taken to ensure that the part of the AI model lifecycle management platform that gets exposed to them doesn't expect them to be data scientists.

An AIOps platform must be set up to learn continuously by using up-to-date data from your environment and to improve based on user feedback. In addition, AIOps products can't be black-box solutions. Companies deploying AI models demand full transparency of the inner workings of our AI models for various reasons, including legal and regulatory concerns. IT operations products employing AI models should be set up to give IT operations administrators access to AI models for triggering retraining for examining model performance on demand, even while having provisions for automatic retraining on a regular basis.

Discrete human handoffs to natural human-AI collaboration

As we discussed in the motivating scenario, we believe that delivering insights where people do their work avoids multiple unnecessary "tool hops" and disruptions for users. Believing in this principle that has been validated with user testing, we deliver insights both in a dashboard as well as in ChatOps environments (such as Slack and Microsoft Teams). Users can switch back and forth seamlessly from dashboards to ChatOps environments without tool hopping.

Compliance by design

We understand that companies need to be able to set their policies and preferences and have the AI and automation honor them. We envision AIOps products to have a flexible policy management framework using which users can specify their policies, rules, and preferences for guiding the AI and insights. For example, if a certain type of event doesn't need to be raised as alerts to users, one can specify those policies in the system. Similarly, certain type of alerts that are self-resolving don't need to be raised up to the level of an incident.

Journey to automation

AI-powered automation doesn't have to be an all-or-nothing phenomenon. Some things can be automated fully; some things may need a human in the loop until trust can be established with automation. Whatever the case may be, we believe having a solid foundation in a platform that supports automation is an essential part of AIOps. Automation functions, include supporting runbook automation, process mining and analysis, and robotic process automation (RPA) are all integral parts of this automation platform. Building on top of such a platform enables us to elevate and connect AIOps insights with the business processes and applications that they monitor and support.

In the rest of the chapter, we give a summary of the AI that we are building into IBM's product, IBM Cloud Pak® for Watson AIOps.

THE AI IN WATSON AIOPS

We are on a journey to realize the vision of AIOps mentioned in this chapter for solving the vexing operations management problems for IT operations engineers. This includes building the various AI analytics described in Table 1. Our journey includes the development of a product called Watson AIOps for bringing what AI can offer to the forefront in optimizing IT operations management.

Figure 3 shows how AI pipelines in Watson AIOps are designed to help SREs in detecting issues early, predicting them before they occur, reducing event and alert noise, locating the specific application or infrastructure component that is the source of the issue, determining the scope of incident impact, and recommending relevant and timely actions. All these analytics help reduce the mean times to detect (MTTD), mean times to identify/isolate (MTTI), and mean times to resolve (MTTR) an incident.

Anomalies are predicted from logs and metrics using log and metric anomaly prediction AI models. The predicted anomalies and other events and alerts that are generated in an IT environment are grouped into their corresponding incident buckets by leveraging various techniques, including entity linking and spatial, temporal, and topological algorithms to reduce event noise. This is done by *Event Grouping AI models*. Faults are diagnosed and localized by *Fault Localization AI models*. The set of impacted components are noted by *Blast Radius AI models*. Similar incidents from the past incident records are identified and next-best-actions are derived by *Incident Similarity AI models*. Finally, problems are avoided by predicting risks associated with deployment and configuration changes via *Change Risk Prediction AI model*. We present a glimpse of how these AI analytics can be realized:

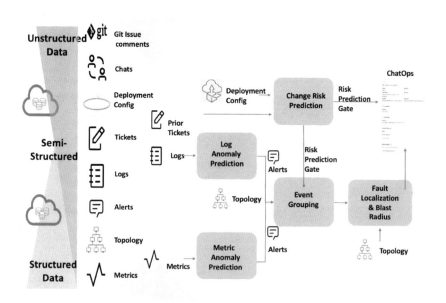

Chapter 7 – Figure 3: AI pipelines in IBM Cloud Pak for Watson AIOps

Log anomaly prediction

An anomaly is something that deviates from normal, standard, or expected behavior. Typically, organizations set either static thresholds or manual rules to define and manage deviations from the normal behavior. The problem with status thresholds is twofold:

It takes a long time for subject matter experts to distill them from their experience and to create them.

They don't adapt to changes and, therefore, tend to get outdated and irrelevant quickly.

If not updated or deleted, these manual rule-based anomalies can start to flood SREs with irrelevant alerts. We use deep learning algorithms to both prepare features from logs during log parsing and to make anomaly predictions. Users don't have to set static thresholds or manual rules to detect anomalies.

Metric anomaly prediction

Watson AIOps' metric-based anomaly detection analyzes metrics data from various systems (e.g., Instana, New Relic, AppDynamics, and SolarWinds) to automatically learn the normal behavior of metrics in your company and detect anomalies from those metrics. It employs a set of time-tested time-series algorithms (e.g., Granger Causality, Robust Bounds, Variant/Invariant, Finite Domain, and Predominant Range) to capture seasonality and significant trends and to perform forecasting.

Event grouping

An event indicates that something that is noteworthy has happened in an IT operations environment. For example, an application has become unavailable or a disk is full/reaching capacity. The goal of event grouping and classification is to reduce the noise for IT operations management personnel and help them focus on a few important events that need their immediate attention. Anomalies detected from metrics, logs, and tickets are grouped using multiple algorithms (e.g., Temporal, Spatial, and Association Rule mining) in Watson AIOps for event grouping.

Static and dynamic topology management

Application and network topology refer to a map or diagram that lays out the connections between different mission-critical applications in an enterprise. Static topology refers to a map that is constructed based on the build and deploys information on applications and infrastructure components. Dynamic topology, on the other hand, refers to a dynamic map that captures the resources and their relationships as the environment changes at run-time and provides a near-real-time visibility of the same.

With Topology Manager in Watson AIOps, you can compare the current topology with a historical one to answer questions such as "what happened?" and "what's happening now?" It helps you investigate the details that led up to an incident and see the topology (and status) changes over time. In addition, faults are localized on topology.

Fault localization and blast radius

Entity mentions are the names of the resources (e.g., service or application component names, server names, server IP addresses, pod IDs, node ID, etc.) that are referenced in anomalous logs, alerts, tickets, and events. Once events are grouped, entity mentions in anomalous logs, metrics, alerts, and events are extracted. These entities are resolved with topological resources to isolate the problem and to place the identified entities on the corresponding dynamic topology instances that match the time at which the mentions were noted. Traversing the topological graph in the application, infrastructure and network layers enable us to map out the impacted components, known as blast radius.

Incident resolution

Watson AIOps ingests and mines prior incident ticket data by connecting to tools like ServiceNow to provide timely and relevant next-best-action recommendations for the currently diagnosed problem at hand. Current incident symptoms are framed as a query to the indexed ticket data to search and retrieve the top k relevant prior incident records and important entity-action (aka noun–verb) phrases are extracted from each relevant record to make it easy for SREs to get a quick glimpse of the suggested action. We apply various NLP techniques to extract entity and action phrases, including rule-based systems.

Insight delivery and action implementation

In Watson AIOps, all of the insights described above are delivered via both ChatOps and dashboards. Real-time, in-the-moment insights are delivered via ChatOps to SREs directly in the place where they work. Within ChatOps, there is functionality to interact and share selected incident resolution suggestions with other collaborators, in addition to exploring the evidence of the insights. From ChatOps, SREs can launch log, metric, and ticket monitoring tools to explore further details. Similarly, SREs can launch interactive dashboards for detailed exploration of events, event groups, metric anomalies, and topology. Applicable actions/runbooks can then be automatically run via Runbook execution.

WHAT'S NEXT FOR AIOPS?

As noted at the start of this chapter, we envision fully instrumented, observable, self-aware, automated, and autonomic IT operations environments in the future. AI can help us get there.

We envision that AIOps solutions will not only be able to help resolve issues in a reactive mode but also help avoid issues from happening in the first place by designing the DevSecOps lifecycle activities for efficient operations right from the get-go. For example, smart checks and gates prevent risky deployments from getting pushed to production, stop under-tested code modules from getting into deployment phases, and block code with risky security vulnerabilities from getting to the deployment phase and so on. Thus, as illustrated in following figure, by instituting feedback and feedforward loops in software development lifecycles, we can develop full end-to-end visibility and manage IT systems better. We can't wait to shape the future and take you all with us in this journey!

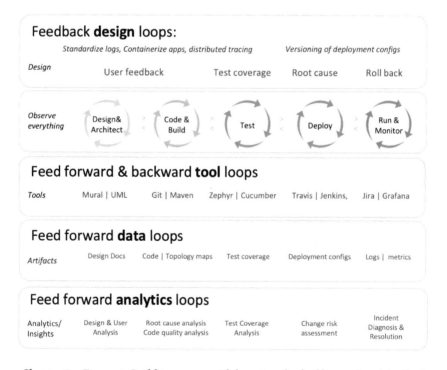

Chapter 7 – Figure 4: Build to manage: Elaborating the feedforward and feedback loops in software development life cycle

UP NEXT

To learn more about AIOps, make sure you check out *The Art of Automation* podcast, especially Episode 1, in which Rama and Jerry discuss automation with AIOps. Rama shares the sentiment of "prevention is better than the cure," further explaining that mission-critical systems, including airlines reservations, financial trading systems, and retail checkout systems, are all expected to be running 100% of the time. They discuss how AIOps changes the strategy of IT operations managers from solely focusing on optimizing "mean time to recovery" to "incident prediction and avoidance." All made possible by AIOps technology.

* * *

The key to a successful AIOps deployment is predicated on the richness and diversity of the data that is emitted from enterprise systems. The on-ramp to this data is APIs.

In 2002, Jeff Bezos professed his now famous decree on internal APIs, which is the topic of the next chapter. It went something like this:

All teams will henceforth expose their data and functionality through APIs… teams must communicate with each other through these interfaces… there will be no other form of internal communication… anyone who does not comply will be fired… have a nice day.[35]

With this, Jeff figured out how to make an online bookstore profitable. Amazon's fulfillment speed and accuracy, through the use of APIs and the automated processes that follow, has been key to its customer success. You see, by mandating that an enterprise *unlock data and automate processes* through self-serve APIs, they significantly remove IT as a bottleneck and empower the entire organization to relentlessly automate internal processes.

So, in the spirit of quotes, Jerry Cuomo has been known to use the following quote: "You can't automate an enterprise unless that enterprise is 'programmable' through APIs." Perhaps this quote is not as dramatic as Jeff's quote, but reading on to the next chapter will illustrate how you can follow this wisdom and make your "online bookstore" epically profitable.

Chapter 7 – Figure 5 – Podcast Episode 1 – Automation and AIOps

Chapter 8

APIs

To build a truly engaging, automated experience, enterprises need to access and bring data together from all systems, applications, and environments. AI to the rescue!

Chapter Author: Rob Nicholson

COVERED IN THIS CHAPTER

- Definition of APIs

- Make enterprises automatable with integration and APIs

- How AI accelerates API creation with "smart field mapping"

- Creating a "closed-loop integration" with AI and automation

WHAT IS AN APPLICATION PROGRAMMING INTERFACE (API)?

While we've already seemingly mentioned the term "API" one hundred times to this point in the book, it seems right to start this chapter with a proper definition. An application programming interface, or API, enables companies to open up their applications' data and functionality to external third-party developers, business partners, and internal departments within their companies. This allows services and products to communicate with each other and leverage each other's data and functionality through a documented interface. Developers don't need to know how an API is implemented; they simply use the interface to communicate with other products and services. API use has surged over the past decade, to the degree that many of the most popular web applications today would not be possible without APIs.

AUTOMATION, INTEGRATION, AND APIS

Most enterprises are driving towards a digital nirvana, where they can deliver an experience to their clients that is both automated and bespoke to the individual. In order to do this, they need to leverage the data from the sum total of every interaction any part of the enterprise has ever had with that user and combine it with any relevant information from outside the enterprise. This data, digested by artificial intelligence (AI) technologies, can then be used to deliver a delightful experience to that individual user.

What this means is that the folks working on this client experience need to be able to access every piece of data they have and potentially interact with any and every back-end system that the enterprise is using. What makes this tricky is that many enterprises today have a mixture of legacy systems that they have been using for years combined with systems that have been brought together via acquisitions and new applications built to support emerging new lines of business. Furthermore, many organizations are partway through a migration to cloud and already are using multiple software-as-a-service (SaaS) applications to replace some of their legacy systems.

So, to build a truly engaging, automated experience, the team needs to access and bring data together from all these systems, applications, and environments to their AI models and then to the user experience layer. Put another way, *integration underpins automation*. This is the integration challenge that IBM is working on with legions of clients around the world.

These days, we typically find that the teams building these new applications are small and agile and deadlines are tight, so the integration team needs the technology to do as much of the work as possible, leaving the human brains free to be creative and think about the really hard problems. Put another way, *we need to accelerate integration using AI*.

Thus, the interaction of automation, AI, and integration can be divided into two focus areas:

1. *Integration as an enabler for automation*: In order to automate the enterprise, the automation must be able to access to all of the data in the enterprise, and it needs to act on that data and on the processes and applications the enterprise uses.

2. *AI as an enabler for integration*: Using AI and automation, integration teams can accelerate the pace at which an enterprise can create the APIs they need.

MAKING THE ENTERPRISE AUTOMATABLE WITH APIS

Let's look at what happens when you automate. Something that is happening in the enterprise is going to trigger an automation to run. It might be an event that occurs or a piece of data that gets generated. The automation needs to be able to access the data or event so that it can know that it needs to take action. Then, of course, the automation has to be able to take actions wherever it needs to, so it needs to be able to reach out and interact with the existing applications and systems in the enterprise.

If you're thinking about automation, you're almost certainly doing it because you want to go faster and/or to free up the human brains in your organization to do more productive work. So, as you do this, you need the people building the automations to have quick and easy access to the data, to the events and the actions that they can perform. The best way to do this is to adopt a strategy to make application programming interfaces (APIs) ubiquitous across your company and to make those APIs as consistent as you can by adopting a common standard and documenting them in the same way.

When we talk about APIs there, I'm not just talking about REST APIs or even SOAP APIs that you might document using OpenAPI or WSDL; we also mean Streams of Events or "Asynchronous APIs" that you might document using a standard like AsyncAPI as well as GraphQL APIs that we're starting to see adopted.

API MANAGEMENT

In order to make usage of APIs ubiquitous, what is needed is an API management practice based on tools that can help you to *Create, Secure, Manage,* and *Socialize* APIs, irrespective of whether the API endpoints are in your on-premises data centers, in the public, or on private cloud(s).

Whichever tool you choose should have powerful tooling to help you *create* high-quality APIs from the endpoints you have. That may involve significant orchestration flows implemented in an application integration tool, or it might simply involve creating an OpenAPI description file and implementing a simple gateway flow.

You will need to *secure* all these new APIs to control which applications and which developers have access to them and even implement policies like rate limiting to prevent runaway applications and automations from taking down your critical backend applications.

You will need to *manage* the lifecycle of your APIs so you can keep control of which versions are being used, publish new versions, and retire old versions.

Just having the APIs is only part of the story. You will also need a way to *socialize* the APIs with your community of automation developers, allowing them to search and find APIs and to understand and access them in a developer portal.

It is all about speed. Each developer building an automation using an automation tool or framework needs to be able to access APIs of all kinds directly from that tool. They need to be able to understand what those APIs do and how they relate to each other. They want to be able to call the APIs and understand the results of the API call. Most important to developers, this tool needs to be a one-stop-shop, allowing the developer to perform all these tasks without ever leaving their automation tool.

IBM API Connect is built to support all these needs, providing state-of-the-art capabilities to create, secure, and manage APIs of all types.

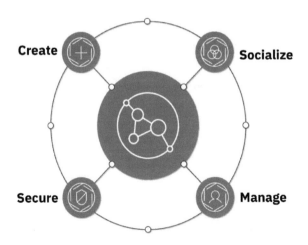

Chapter 8 – Figure 1 – IBM API Connect Overview

ACCELERATING INTEGRATION USING
ARTIFICIAL INTELLIGENCE

There is no end to the appetite for integration in a modern business. Every central IT team has a backlog of integration projects many times longer than their capacity to deliver. New business imperatives insert high priority projects at the "head of the queue," leaving a very long tail of valuable projects that will never be realized. In response, frustrated line-of-business (LOB) executives are bypassing central IT and spinning up their own "shadow IT" projects.

We think there are three solutions to this problem:

- Enable integration developers to develop new integrations at many times the speed they were previously able to do by equipping them with powerful intuitive AI-based tools.

- Enable many more individuals to develop high-quality integrations by creating tooling that allows non-specialists to leverage expertise built into the tools.

- Integrate these tools into a platform, including specialist tools and API management, to allow central IT to retain control over integrations that are built by shadow LOB developers.

Powerful integration tools that build in expertise

We discussed earlier that the key to making the enterprise automatable is to expose all of the capabilities as APIs, yet the task of creating large number of high-quality APIs is a potentially daunting one, requiring legions of skilled integration developers. A quick search in any online bookstore will elicit dozens of books running to thousands of pages on the subject of designing high quality REST APIs. The question, then, is how can we enable engineers who don't have the time to spend weeks studying to create high-quality API integrations?

The approach that we take in App Connect—part of the IBM Automation platform—is to build the expertise into the tool. App Connect adopts an opinionated view about the shape of an API based on established best practices. Rather than starting the journey by asking the user to create an OpenAPI definition, App Connect starts by asking for the definition of the data model, as shown in the figure below.

CarRepairClaim	✅ Properties	⋯ Operations
Add properties to your CarRepairClaim model		
Name	String	⌄
eMail	String	⌄
LicensePlate	String	⌄
DescriptionOfDamage	String	⌄
PhotoOfCar	String	⌄
CaseReference	String	⌄
ContactID	String	⌄
EstimatedDays	String	⌄
EstimagedBill	String	⌄
Add property +		

Chapter 8 – Figure 2 – Defining an API Data Model

A data model typically supports four basic functions: Create, Read, Update, and Delete (CRUD). In a similar convention, an API model is typically defined by RESTful actions, specifically the corresponding HTTP methods of POST, GET, PUT, and DELETE. With App Connect, the user does not need to know REST; they just select the appropriate CRUD operation and App Connect knows how to map this onto the appropriate RESTful HTTP method.

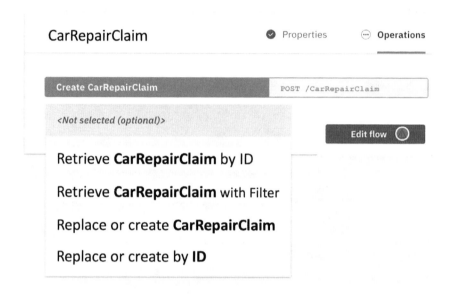

Chapter 8 – Figure 3 – Mapping to RESTful HTTP method

Then, the flow can be built up by selecting the appropriate connectors for the applications and endpoints they need to integrate with. In the past, this would have required the specialist integration engineer to study the documentation for the application to understand its data model and how its APIs were structured in order to figure out how to extract the data they needed.

App Connect skips this time-consuming step by building this information into the connector itself. The IBM development teams did all the work to understand the application and map it onto the consistent CRUD model that App Connect uses and built this into the connectors to be smart connectors. For example, the below figure shows a user searching in Salesforce for contacts with email addresses that match the submitted address.

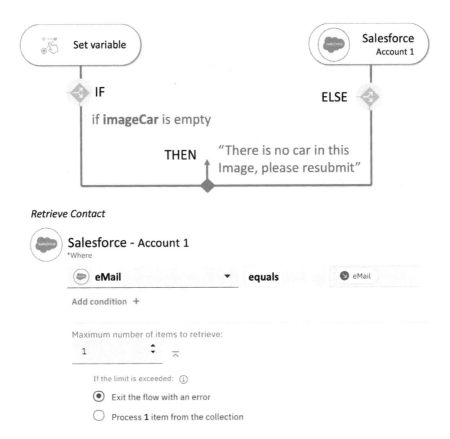

Chapter 8 – Figure 4 – Mapping to Salesforce API with matching email address

The simplification that App Connect brings by rendering every application that it connects to in the same way significantly accelerates work, especially for shadow IT professionals who perhaps do not spend every hour of every day creating integrations and are not steeped in all of the intricacies of each and every SaaS and bespoke application that their enterprise uses. Wherever possible, App Connect provides an interface that uses the objects that the users will be familiar with from using the application. For example, when adding a "card" in the productivity application Trello, Trello API requires the "board_id" and "list_id" for the list to add the card to. That means that an integrator needs to understand what those IDs are and how to find the IDs for the particular board and list they want to add the card to. Rather than the integration developer figuring this out, it's much more useful for the connector developer to build this knowledge into the connector, so that App Connect can allow the user to pick the human readable board name and list name that they recognize from the Trello UI rather than the IDs that the API needs.

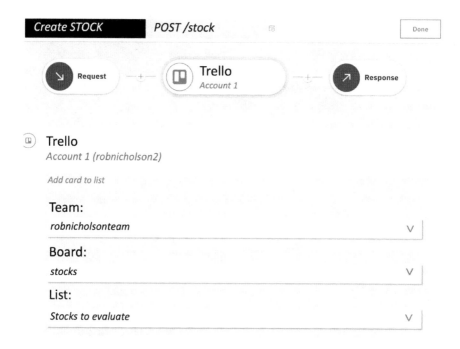

Chapter 8 – Figure 5 – Auto recognizing API fields

Event-driven automations

One of the most common automation scenarios is "When this event happens, do this action or set of actions." Integration folks will recognize that as an event-driven integration flow. Using the same tooling that we use to create API flows as described above, we can create automations that are event driven. For example, suppose we want to send an email to all new sales leads that get entered into Salesforce, but we know that sometimes the leads get added without an email address, we might create an automation flow, like the one below, that is triggered whenever a new lead object is created in Salesforce. If the email field is not filled in, it sends a Slack message to the team asking them to reach out by phone, otherwise it sends an email. Building a flow like this would be complex without smart connectors that can trigger event-driven flows, but by using the expertise built into App Connect, a flow like this can be built in a few minutes:

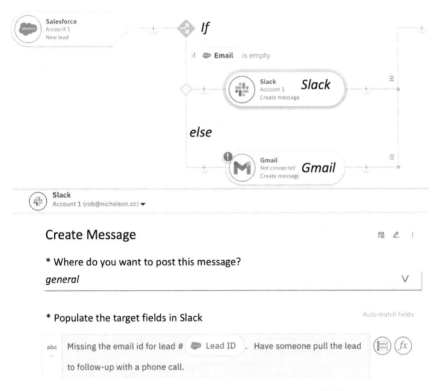

Chapter 8 – Figure 6 – Auto recognizing API fields

AI-based "Smart Mapper"

We asked our customers what activity their integration developers spend the most time on. The answer that came back resoundingly was "mapping data"—specifically converting the fields and data representations in one API or application to the fields and data representations in another. Spurred on by that, we developed App Connect Smart Mapper. We think of this as an assistant to the integration user. It does not take over completely, but it can lighten the load considerably.

It uses a range of natural language processing techniques, including machine learning models and rules, to calculate the most likely mappings to a set of target fields from all of the data that's in context at a given point in an integration flow. Like all good AI assistants, it is driven by confidence levels. The user can choose to have the tool insert all of the highest confidence mappings or just a subset. Even for mappings that the AI is less sure of, it can provide assistance, narrowing down the mapping task from thousands of possible fields to a small number of suggestions with confidence levels, allowing the user to make selections.

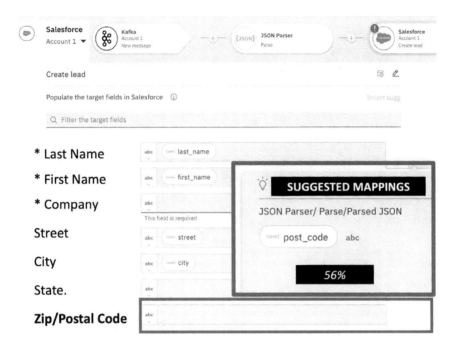

Chapter 8 – Figure 7 – AI-assisted API field mapping

Although clients have told us already how much time this saves them, what we have so far is only the start. In the lab, we have online learning models that can adapt the mapping suggestions made based on data from previous mappings. Not simply learning particular mapping that can be repeated again verbatim but building up concept associations that allow the AI model to make suggestions in situations they have never seen before based on the concepts that they have learned from previous mappings. This online learning creates a feedback loop that we call "closed-loop integration." The section that follows describes how this style integration uses operational and development time data, feeding it back through machine learning models to automate and accelerate integrations.

CLOSED-LOOP INTEGRATION

This theme of "closed-loop integration"[36] is an important and recurring concept in integration. By taking real-world, company-specific operational data and feeding it through AI/ML models, we can use it to make recommendations and optimizations that make integrations more accurate, faster, and more robust.

One example of this is the API testing feature of API Connect. This feature addresses one of the most significant pain-points from our clients, that of developing a comprehensive suite of tests to validate the correct operation of APIs and to ensure that they do not break as they are enhanced and versioned. The API test feature can automate the generation of a suite of tests based on the OpenAPI specification for an API. It can even generate tests that establish a particular application state within the API implementation, necessary to execute the full range of the API's capabilities.

Following the principle of closed loop integration, we have AI models from IBM Research that take Open-Tracing data from production deployments and use this to calculate the behavioral coverage that the current test suite produces. Where coverage gaps are detected, the model can suggest new tests that can be generated to enhance test coverage. Thus, feeding back operational data allows us to improve the quality of the resulting API and accelerate the development of new versions of the API because the API developers can proceed at speed knowing that they have a thorough test suite to guard against regressions.

UP NEXT

We have seen how integration underpins automation by allowing an enterprise to quickly access and automate all the data and processes. We have also seen that in order for a modern enterprise to achieve this Herculean task, it must leverage powerful tools that provide a dramatically simplified user experience, leveraging AI to enable many more of its employees to get involved with creating integrations.

Make sure you check out *The Art of Automation* podcast, especially Episode 11, in which Rob and Jerry discuss automation and APIs. They discuss the relationship between APIs, integration, and automation and why "you can't automate an enterprise, unless that enterprise is programmable through APIs." Rob describes how, in order for an enterprise to grow faster, it needs quick and easy access to all relevant data and events, and how the best way to do this is to make APIs ubiquitous across the company. He also shares examples of how AI can be used as your "wing person" in tough integration problems.

* * *

Now that we've read our expert's views on the fundamentals of both business and IT automation, it's time to hear from a few industry luminaries. In the next set of chapters to follow, the book will change its format, switching to an interview style. These chapters focus on automation in the context of specific industry settings. Jerry conducts interviews with leaders in the *healthcare, insurance, retail, and financial services* industries. We also tacked-on a few additional interviews that exhibit out-of-the-box thinking on the very interesting topics of *automation and the weather* and *automation at sea*. Read on to hear firsthand how AI-powered automation is making a material difference of the everyday life of both the workers and customers across these industries.

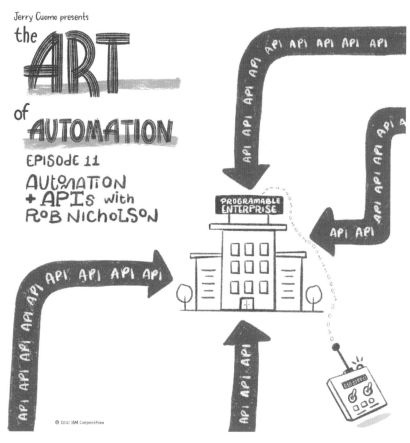

Chapter 8 – Figure 8 – Podcast Episode 11 – Automation and APIs

Chapter 9
HEALTHCARE

A conversation between Jerry Cuomo and Claus Jensen from the Memorial Sloan Kettering Cancer Center on automation in the healthcare industry.

Interview by: Jerry Cuomo

COVERED IN THIS CHAPTER

- Automation in the healthcare world

- AI in the healthcare world

- A hybrid care model

- Moving to the "next normal"

AN INDUSTRY LENS

This chapter of *The Art of Automation* is a reduced transcript of a conversation between Jerry Cuomo and Claus Jensen, chief digital officer at Memorial Sloan Kettering Cancer Center and author of four *For Dummies* books. In our first chapter of AI-powered automation viewed through the lens of a specific industry, Claus and Jerry discuss the important role automation is playing in healthcare and why the combination of artificial intelligence (AI) and human doctors outperforms either in isolation. They also explore both good and bad examples of how AI can be used in healthcare and what innovations they expect to see in the next decade.

JERRY AND CLAUS

CUOMO: Welcome to *The Art of Automation*, a podcast that explores the application of automation in the enterprise. I like to start these podcasts with a simple definition of automation, so here it goes. Automation uses technology to automate tasks that once required humans. Simple enough, right? But, as you'll see, the devil's in the details, and it's those details that we like to bring to the forefront with these podcasts.

So, let me tell you what's in store for today. Today we're going to take our first up-close look at automation from an industry lens, and specifically, healthcare. What better guest to join me for this discussion than Claus Jensen?

Claus is the chief digital officer at Memorial Sloan Kettering (MSK), where he's leading digital business transformation, innovation and platform initiatives, and the overall technology agenda. At MSK, Claus is focused on scaling and expanding the fight against cancer through a seamless synthesis of extraordinary people and meaningful technology—I really like that!

Before Sloan Kettering, Claus served as the CTO and head of architecture for CVS Health and Aetna. But wait, there's more. Claus is also the author of four *For Dummies* books by Wiley Publishing (John Wiley & Sons, Inc.), including *Digital Transformation for Dummies*, *The Hybrid Cloud Journey for Dummies*, and *APIs for Dummies*.

And with that, I'd like to welcome Claus to *The Art of Automation*. Welcome, Claus.

JENSEN: Thanks, Jerry.

CUOMO: Great to have you. And look, I've chatted with you a number of times, and I know you have so many good perspectives here, so I just want to jump right into it if you don't mind.

JENSEN: I don't mind at all.

CUOMO: Great.

AUTOMATION IN THE HEALTHCARE WORLD

CUOMO: Claus, what do you see the role of automation being in the healthcare industry; and, why now?

JENSEN: So, the first thing to say is it won't replace the doctors. I mean, that is sort of what you see in popular literature, that this is going to replace everything. But it is going to make, hopefully, healthcare a better experience.

We've all seen the discombobulated nature of things not being connected. There's a lot of friction in healthcare, and we have seen it very close up over the last seven months dealing with an unforeseen pandemic.

The real question becomes, how do you use automation to reduce the friction—give everyone a better experience? And not just the patient and caregiver side, but this notion of burnout on the provider side is quite real. What burns people out is more about the meaningless stuff that you have to do because that's just how the systems work; it's not necessarily about the volume of work that you have to do. We all get motivated by the things that we care about and we all get slightly demotivated by red tape and stuff we have to do "just because." So, I'd say that's probably the biggest role of automation in healthcare.

CUOMO: So, not replacing the doctors but certainly enhancing the experience of all involved.

JENSEN: Yes.

AI IN THE HEALTHCARE WORLD

CUOMO: Claus, could you explain the role of artificial intelligence in automation, and if you can, maybe give a couple examples?

JENSEN: So, let's start with this debate in the industries of which is better—is the doctor better, is AI better? It's an interesting question and actually doesn't matter, because what we do know with certainty is that the two together are better than either in isolation. We've seen that.

Any time you do a study, you can debate which one is better depending on the problem, but if you put the two and combine them in the right way, it will inevitably end up with higher quality and a better experience. I'll give you a good example and a bad example.

The good example is if you take a picture of tissue to a cryo-electron microscopy. If you've ever seen one of those images, it looks a little blurry and you can barely discern any structure by the naked eye. But an AI model can, and it can enhance the structure, and it can highlight what the structure looks like, and I can have a human oncologist—a human physician—actually interpret. So what does it mean in the clinical sense? That's a great example of AI being used to enhance something we can't see with our eye.

A not-so-good example was a hospital—wasn't ours—that was training an AI model to try to recognize disease, and they wanted to do the best possible job so they gave this training algorithm all the images from their best piece of equipment. And it worked great in tests, and they got to production and the results didn't make any sense.

What's going on? It turned out they had trained their AI algorithm to recognize images from that piece of equipment, which happened to also be the piece of equipment they sent their most sick patients to go get diagnosis from. It was an unforeseen bias in the data and the way you had trained it. So, you've got to be careful. If you're not careful with how you train it and how you feed the machine learning algorithm, you'll get an unexpected result.

A HYBRID CARE MODEL

CUOMO: Claus, you have a number of responsibilities in your role as the chief digital officer across IT, technology, business, and more. What would you say is the biggest help technologically over the last couple of years to bring your team together? You just gave an example of computer vision; any other examples of particular technologies that you see propelling automation?

JENSEN: I think there are three, and they're tied to three vectors of change around us. There is the whole move to a hybrid care model—so the emergence of not just audio/video telemedicine tools but perhaps more importantly, our ability to stay in touch in a near real-time fashion when you're not necessarily in our facility. I can send people home with devices and I can get real-time telemetry into my offices and basically take care of them as if they were in my hospital.

The second one is we've gotten new tools and technologies to understand data semantically and actually try to make sense out of a very rich data set that we have historically not been able to make sense out of, partially because a lot of it is written notes from physicians, and they don't necessarily speak in structured data—they're not supposed to—to tell a story about what happened with the patient.

And then the third one—and that's actually where I think we will see the biggest wow factor if you sort of look ahead—is much of the healthcare system is focused on disease, and we know how to take care of disease. But the real question is, how do we help people that haven't gotten sick yet? And that's not a problem that you can scale with people and physical resources, because there are too many humans in the world and you can't call everyone every day. But you could build a digital platform that allowed people to get guidance and knowledge in a pre-disease stage, and then you can integrate in humans with specialized experiences and skills when you need it. But, by and large, this is a primarily digital relationship that helps you with all of your needs logistically and practically and knowledge-wise around some particular disease.

And we haven't seen that yet—not really. But I think we finally have the technology to build it; and if we could, this will be a lot smarter than Google and a lot more useful than any static library you can go read in a book, because it will be personal, it will be meaningful, it will be about you.

MOVING TO THE "NEXT NORMAL"

CUOMO: When you think about moving forward past this pandemic, all eyes are going to be on industries and how we move forward, get people back to work and optimize our overall surroundings. What's your view on how we get back to our new normal?

JENSEN: It's a great question, and I don't even know whether I think that the "new normal" is the right term. I sometimes talk about the "next normal" because, in some ways, it's about reinventing what does normal even look like. True transformation is almost inevitably based on both optimization and creation. You need to optimize some of the stuff you have to free up resources to create new things, and we talked a little bit about some of the new things I would like to bring into the world of cancer.

Usually, when you talk about optimization, people tend to say, how can you do more with less, right? So, if I give you less of the resources that you have today, can you do more with the same—with a smaller amount of resources—and get the same piece of work done?

I actually think that's the wrong question. I think the right question is, how do we do more with more? And here's one—because as we get more capabilities in terms of digital and intelligent technology, we have more tools in the toolkit. There are more problems we can solve; and as we solve more problems, we can actually make some of the existing problems smaller.

I'll give you a simple example. If you can teach people how to manage risk earlier in the lifecycle, you get less need for some of the heavy lifting that comes further downstream when they get sick. It's a simple example that's related to population health management, but it's the same principle.

When you have got more tools in your toolbox, I think we should look for doing more things with more tools—hence, getting to a better outcome—rather than just asking ourselves, can we do the same things with less resources. That's an uninteresting question.

CUOMO: Claus, we spoke earlier this year—I think I'm going to guess it was around January—and you made a statement which, I've got to admit, when I heard it, I didn't think of it too much, but when I took the role in automation, I thought back right away to you.

Before the pandemic, even, you had said that 2020 increasingly will benefit from automation across the enterprise, but in 2021-2022 it's going to be a really big deal. And I'm paraphrasing what you said, but you kind of like foreshadowed the importance of automation. And jeez, like look what's happened in between. I don't know that it's a year that any one of us expected.

So, can you pinpoint one or two particular areas where you feel excited about automation making everyday life for you, your staff, your patients, people yet to be diagnosed, as you said, better?

JENSEN: Well, if I were to pick one, it is the ability to have a seamlessly orchestrated process all the way from identification of I need to meet with someone, I need to go through a whole bunch of tests, and ultimately ends up with a disposition after the visit. In the old world, this was a very cumbersome process. It involved a lot of people and a lot of paper.

In the world we now live in, you can imagine a CRM-type tool, and it will allow you to do lightweight orchestration of the process. You can plug in people that are not clinicians. You can orchestrate the logistics, and, ultimately, all you have to do as a patient or as a physician is your part of that process instead of having to worry about all the red tape.

In some ways, this is the most low-hanging fruit but also the most profound change for any of us who live in the healthcare environment, because all of a sudden, we just have to worry about our part.

UP NEXT

In the interview, Claus talks about the need to "make sense out of a very rich data set that we have historically not been able to make sense out of." In healthcare organizations, many repetitive processes and decisions rely on the availability of accurate data. For example, patient onboarding and follow-ups, medical billing and claims processing, generating reports for physicians, and prescription management are some of the repetitive tasks that are common across all healthcare organizations. These are among the factors that create the perfect environment for RPA to increase efficiency, reduce costs, and improve patient experience. To learn more about RPA's use in healthcare, read the article by Paula Williams, IBM RPA – "Amazing Ways That RPA Can Be Used in HealthCare".[37]

* * *

Make sure you also check out *The Art of Automation* podcast, especially Episode 3, from which this chapter came. Listen in to the full episode to hear Jerry and Claus discuss the important role automation is playing in the healthcare industry.

Chapter 9 – Figure 1 – Podcast Episode 3 – Automation and healthcare

Chapter 10
INSURANCE

A conversation between Jerry Cuomo and Carol Poulsen from The Co-operators Group on automation in the insurance industry.

Interview by: Jerry Cuomo

COVERED IN THIS CHAPTER

- Disruption in the insurance industry

- Automation in the insurance world

- AI in insurance

- Shifting risk and lowering costs

ANOTHER INDUSTRY LENS

This chapter of *The Art of Automation* is a reduced transcript of a conversation between Jerry Cuomo and Carol Poulsen, chief information officer at The Co-operators Group. In our second chapter of AI-powered automation viewed through the lens of a specific industry, Carol and Jerry discuss the important role automation is playing in insurance and how companies are moving from standard risk mitigation to artificial intelligence (AI)-powered risk prevention. They also explore how automation is allowing insurance companies to put customers at the center, instead of at the end of a process. Finally, they predict what AI innovations could come to insurance in the coming years and how those innovations could shift risk and save you money.

JERRY AND CAROL

CUOMO: Welcome to *The Art of Automation*, a podcast that explores the application of automation in the enterprise. You all know I love talking about autonomous vehicles and using them as an example of how AI and automation can team up with us humans to make everyday life better; and in this case, I believe, safer.

Well, such advancements in AI-powered automation are causing a ripple effect across multiple industries; and specifically, think about how these vehicles might disrupt the landscape of the insurance industry. It's entirely possible that auto insurance will change into something very different than what it is today.

Think about it—autonomous features are supposed to make cars safer, right? Well, you would think this would result in lower car-insurance premiums; or, would they result instead in higher premiums because when crashes do occur, the high-tech features make repairs extremely expensive? How does an insurer actually calculate risk? Who are the parties that are liable if an accident does occur?

Well, with the introduction of smart devices in the home, on the road, and on your wrists, we are seeing a glimpse of the fuel by which the insurance industry can reinvent itself into a data-driven AI-powered automation engine that is poised to make everyday life better, safer, and maybe even more economical.

Today's guest has a proven track record of putting emerging technology to use to transform industries; and specifically, who better to discuss automation in insurance than my friend Carol Poulsen, who is the CIO at The Cooperators Group? As you all know, Cooperators is a leading Canadian-owned insurer offering auto, home, life, farm, travel, and business insurance, as well as investments.

Now, before Cooperators, Carol held several leadership roles at TD Bank and the Royal Bank of Canada. Her motto is "disrupt or be disrupted," and this is where we'll start our discussion. Welcome, Carol, to *The Art of Automation*.

POULSEN: And here we are again, Jerry.

INSURANCE CAN BE EXCITING, TOO

CUOMO: Carol, can you share with our listeners why you're so excited about the insurance industry; and, specifically, why now? And think about the current events and how they're putting an exclamation point around insurance.

POULSEN: Sure, I'd be happy to. I think what I find really exciting about the insurance industry right now is we are doing our best to disrupt our own business. I really think that that's the way forward. If we don't, somebody will disrupt it for us absolutely.

The ways in which we're trying to disrupt ourselves center around two aspects. The first one is the traditional relationship of an insurer to a client, and we're trying to put the client at the center as opposed to the end of a process.

The way in which we're doing that is to look at what they need from us. History has been that we have provided risk mitigation. So, you buy an auto policy, a home policy, or a life policy, and our relationship with you is you pay money every month for these policies and, hopefully, nothing happens and you never hear from us. But, if at some point, your house burns down, you get in a car accident or, God forbid, somebody dies, you really hope that we treat you well.

CUOMO: Right.

POULSEN: That's really not the kind of relationship we think the future demands, and that when we think about risk mitigation to our clients, it's more than mitigation after the risk has happened. It's also about risk prevention.

We're very focused, right now, on looking at ways through AI, through automation and through relationships with other companies in other industries that we can absolutely provide prevention advice and prevention services. So, you might think IoT (Internet of Things) in the home, on the body. You might think that we would work with large companies that have alarm services that we could offer blended service to and really be out there in front of the event happening.

We started this actually before we started trying to disrupt ourselves with something as simple as a text letting you know a hailstorm's coming. You could put your car in the garage. You're much happier and so are we.

CUOMO: Useful, yes.

POULSEN: So, that's just a tiny way of looking at it, but what we really want to do is wrap risk mitigation around property and casualty insurance, around life insurance, so that we're all more focused on preventing the risk than paying out after something unpleasant has happened.

The other thing that I would just mention is the changing face of transportation, too. And to say that it's a broad term, but it means everything from automated vehicles and who is the insured then and who is at fault if an accident happens, to the fact that many car manufacturers are starting to offer a bundled insurance offering when you buy the car, which is changing the point of origination. We need to be there with the car companies when this happens.

So, again, that's a huge disruption for the insurance industry that provides an interesting challenge and an opportunity.

AUTOMATION IN THE INSURANCE WORLD

CUOMO: Carol, this is *The Art of Automation*, so let's talk about the relationship between insurance and automation.

POULSEN: Well, if you think about it, automation has been around for a really long time. What's different now with insurance is automation can be offered at so many levels, and there are so many new tools and services and products out there for us to use to automate. I will go everywhere from a basic sort of pedestrian robotic process automation (RPA), which is really about looking at where is there a process happening that's very manual, and we're going to stick an RPA process in there instead. I really view that as a BandAid because, ultimately, you're going to have a bunch of these and you're going to refactor all of that at some point.

CUOMO: Most likely true, yes.

POULSEN: So, I view that as a time-sensitive solution but one that's working really well. We've business cased all of those, and we're finding great value in them, at present, and we'll keep using them because you're not transforming everything simultaneously. There's always some area that you're not working on right now that might benefit by some RPA.

A sort of middle tier for me of innovation and automation would be a chatbot. And, as it turns out, we have a chatbot in our life insurance area that is there to advise insurance advisors on some of the more complicated aspects of life insurance. We happen to have powered it by Watson-as-a-Service through the IBM Cloud. And it's doing very, very well.

And then really much larger views on automation are bringing in tools to automate a service that may be between ourselves and another company. It may be multiple companies working together, and we're automating an offering that hasn't existed before. I also think about automation in terms of our data center services and using AI to get insights into those so that we're offering cleaner, more robust services. I also think about it in terms of claims settlements, for instance. We are using some new tools in claim settlements so that we can deliver straight-through processing for claims. So, it's everywhere.

CUOMO: Any sense of the efficiencies you get as you add automation?

POULSEN: I would say, first of all, if you're replacing a person, automation doesn't need to sleep—it doesn't need to take weekends or holidays—and it is far more consistent than a human being, as well. So, not only are you getting 7/24, 365 out of that piece of automation, if you wish, it's also doing it remarkably the same, every time. There's a huge advantage there.

If you're automating something like a claims process to be straight through, this process is going to save you tens, hundreds of people depending on the size of your organization and the amounts of the claims that you're automating to go straight through. So, it scales, and I think it's how you deploy it and how you think to apply it that is the limiting factor.

AI IN THE INSURANCE WORLD

CUOMO: Okay, Carol. Let's join into this party of insurance and automation, let's add the AI ingredient. You had mentioned just briefly some application of AI; can you elaborate a little bit about the role of AI in automating in the field of insurance?

POULSEN: Sure. Insurance processes almost always require dealing with a lot of data. In the past, that has been manual, and that's why you have underwriters and claims adjusters looking at huge amounts of data. If we want to ensure that our services can be digitized so that people can access them when they need them, if you want to ensure that we are as rigorous as we can be in our ability to offer service in a price-conscious way, you're going to want to automate so that you are able to offer those services in a digital fashion. And if it's automated, you can then put in the insight that you're going to gain from data.

And then what data am I looking at? A person can only look at so much data at any given point in time. What we've had traditionally in insurance—and I have to say insurance isn't quite up there with, say, retail in terms of usage, but that's why we're trying to disrupt ourselves to move faster into this space. So, if we're going to be looking at data to get insights to make the best next action—whether it's solving a claim or deciding whether to take on a risk in the underwriting process—well, we can look at all kinds of data.

We can look at data that is not organized. We can look at it from any source, and we can look at it between partners if we start building relationships with other companies and other businesses that are aligned to ours but not the same.

SHIFTING RISK AND LOWERING COSTS

CUOMO: Can you help put it all together? Can you share a view of the art of the possible of these stars aligning? You were touching on some of it, but maybe another example that illustrates where this all might be going.

POULSEN: Okay, let's say that I'm going to work with a company that offers security services in your home. And let's say between us, maybe you get a discount on your insurance because you have these security services. Let's say the company that offers the security services and we share data that we're allowed to share on the risk and on feeds from all of the IoT devices that are now in your home about the risk.

What if we are able to stop a fire from actually getting to the point of an issue with your home? What if we are able to look at your risk entirely differently because of the data that we're getting in and it starts to tell us that we can shift the costs because of the data, we now have that supported by data that we didn't have access to before.

CUOMO: I see, yes. Very cool. I want that service! I actually don't mind sharing data if I'm getting value like that back. Prevention is a big deal, and I think the way you're outlining it, we have all the pieces; I think now you need the insights and the analytics to go with it. And you're right, now that I think about it, who better to do that than you in the insurance world?

POULSEN: Yes. Insurance is really all about understanding a risk and understanding what it costs to mitigate that risk.

And I'll toss you another example with cars. Let's say we worked with a car manufacturer, because they are now collecting all kinds of data as your car is in motion with you in it—everything from the wear on your tires to the wear on your brakes to what radio station you might be listening to in your car.

So, let's say we worked with a car manufacturer and we shared the data—again, that we're allowed to share—and we're able to make recommendations to you about your driving that would allow us to lower your insurance and would allow your car to last longer than it is, like you can get benefits from both vendors in this case.

CUOMO: Yes, I'll order one of those, too. All right, Carol. Thank you very much for joining us here at *The Art of Automation*, and we really appreciate your insights and I hope to have you back again sometime soon.

POULSEN: It was great to chat, Jerry. Thanks so much.

UP NEXT

After listening to Episode 12 of *The Art of Automation* podcast on this very topic, a listener asked if Jerry could provide a comprehensive list of "the sorts of things" that could be automated in an insurance industry context. Working with colleagues from IBM Consulting, an extensive list was produced, and we thought it would be interesting to share that list here. The listing also attempts to estimate the "typical labor savings" for each category in the list. The following figure is a snapshot of that list, which can be summarized by the statement "just about everything and anything can be automated"—a sentiment that Carol alluded to above.

RISK MANAGEMENT | UNDERWRITING	CUSTOMER INTERACTIONS
• Reinsurance Management	• Policyholder Accounts
• Underwriting Management	• Correspondence Handling
• Underwriting Decisioning	
• Health Evaluation	
• **Typical Labor Savings: 20%**	• **Typical Labor Savings: 25%**
BUSINESS ADMINISTRATION	**CASH FLOW MANAGEMENT**
• Procurement/Vendor Management	• Reconcile Cash Transfers
• Business Process Management	• Billing / Collection
• Asset Management	• Payments / Funds Settlement
• Technology Management	• Loans & Dividends
• Auditing/Legal/Regulatory	• Financial Reporting
• **Typical Labor Savings: 25%**	• **Typical Labor Savings: 25%**
POLICY ADMINISTRATION	**BENEFIT & CLAIMS ADMINISTATION**
• Policy Administration	• Claims Management
• Contract Settlement Agreement	• Fraud Management
• Content Management	• Claim Impact/Expert Evaluation
• Document Print and Image Services	• Benefits/Claims Processing
	• Claims Negotiation & Settlement
• **Typical Labor Savings: 30%**	• **Typical Labor Savings: 35%**
BUSINESS ACQUISITION	**PRODUCT MANAGEMENT**
• Campaign Management	• Market Research & Analytics
• Campaign Execution	• Product Deployment
	• Promotion and Brand Management
• **Typical Labor Savings: 40%**	• **Typical Labor Savings: 10%**

Chapter 10 – Figure 1 – List of automatable tasks for insurance

Make sure you check out *The Art of Automation* podcast, especially Episode 12, from which this chapter came. Listen in to hear Carol and Jerry discuss the important role automation is playing in the insurance industry and how companies are moving from standard risk mitigation to AI-powered risk prevention.

Chapter 10 – Figure 2 – Podcast Episode 12 – Automation and insurance

Chapter 11

RETAIL

A conversation between Jerry Cuomo and Tim Vanderham from NCR
Corporation on automation in the retail industry.

Interview by: Jerry Cuomo

COVERED IN THIS CHAPTER

- Automation in the retail world

- Retail on the cutting "edge"

- AI in the retail world

- Crypto in retail

AN INDUSTRY LENS

This chapter of *The Art of Automation* is a reduced transcript of a conversation between Jerry Cuomo and Tim Vanderham, senior vice president and chief technology officer at NCR Corporation. In this chapter
of AI-powered automation viewed through the lens of a specific industry, Tim and Jerry discuss how progressive retailers are beginning to use

automation to transform customers' shopping experience—a transformation that has been catalyzed by the ongoing pandemic.

Tim shares automation examples from self-checkout to online ordering and explains how this goes beyond classic automation to edge, computer vision, IoT, and even cryptocurrency. They close by jumping the tracks to artificial intelligence (AI) and diving into how, out of necessity, retail companies are utilizing AI to filter through millions of events to isolate outages and create a more personalized experience for their consumers.

JERRY AND TIM

CUOMO: Welcome, everyone, to *The Art of Automation*, a podcast that explores the application of automation in the enterprise. Folks, today's "always-on" world has led to new rules for engaging consumers. As shoppers' needs and demands change, retailers need to respond accordingly and quickly.

Consumers are increasingly expecting efficient, safe, and engaging online experiences to also be replicated in store. And as you'll hear from my guest today, progressive retailers are using automation and AI to transform the retail experience—making sense of immense volumes of data, filtering out the noise from millions of events generated in and above the retail store. They're using advanced technology like edge computing, Internet of Things, computer vision, hybrid cloud, and even cryptocurrency.

So, for today's episode, we have the fortune to have my good friend Tim Vanderham with us. Tim is the senior VP and chief technology officer at NCR Corporation, leading NCR's global software and technology organization.

His team includes software innovation and software engineering, with more than three thousand engineers building NCR's current and next-generation products and solutions. With that, I'd like to welcome Tim to *The Art of Automation.*

VANDERHAM: Hey, Jerry, thanks for having me. It's great to be here with you today.

AUTOMATION IN THE RETAIL WORLD

CUOMO: I'm so excited, Tim. I want to get right into it. Why are you so excited about the opportunities to transform the retail space; and, why now? What is happening in the world today, and why is that putting an exclamation point on all of this?

VANDERHAM: Yes. Here at NCR, the passion of automating retail is really out of necessity. When you think about what retailers are going through in the world today with rising labor costs, the challenge they're having of keeping their shelves stocked with the demands that are coming to people in the store, above store ordering, etc., automation is the way they're going to still make the bottom line and still serve all of us as consumers.

And COVID accelerated it. We saw this as a necessary coming down the pipeline—we were planning for it. But COVID hit March of 2020, and bam, we're in the middle of "how do we better automate the retail experience and the retail store?"

I'm excited about it, my team's excited about it. And more than being excited about it, our customers are demanding it. They're coming to us asking:

- How can we deploy software solutions faster?

- How can we automate inventory management?

- How can we automate that whole notion of "I order online, I pick up in the store"?

- How do we automate that flow end to end?

Now think about it in the convenience store space, think about it in a department store like Macy's. It's happening in spades across all of our retailers, across the globe. It's exciting.

CUOMO: Yes, I can imagine because I'm a consumer too, so I see what's happening and really appreciate it. But tell us a little bit more about some of the processes that you're focused on. NCR is known for the cash register, but you're so much more than that these days. Share with us some of the processes that are working behind the scenes that are targets for automation.

VANDERHAM: Sure. So, how are we running the store more efficiently? It's everything from self-checkout, which we have market share leadership in when it comes to self-checkout machines across the globe. But also then how do you just manage all of those components that run in your store?

When you go into a Whole Foods (let's use as an example, a great customer of ours), every self-checkout lane or every manned checkout lane is literally running a computer today, right? Running a piece of software that you interact with as a cashier or as a consumer in self-checkout. How do we keep that up and running? How do we keep that at an SLA level available every day, every hour of that day? So, in-store automation is really key.

We automate things in the store, but then we're using IoT technology to be able to get alerts off of it so that if one of the processes goes down or a printer goes offline we can restart, through advanced analytics and automation—restart that printer and/or restart that process so that manned lane or that self-checkout lane is only down for a couple of minutes. Versus having a store manager call, somebody log in, somebody to drive a truck to fix a printer, whatever it might be. So, that's kind of one example of automation in the store.

CUOMO: Yes, okay. Tim, just give me a sense here—I can imagine years ago where the cash register was a physical device that had probably firmware that got updated infrequently. What is your frequency of change? Is it measured in weeks, days, or finer grained than that? How often are you automating changes at that level?

VANDERHAM: Yes, so, it has been increased. It's not where it needs to be yet today—I hate to say it. I manage everything from code that's been built in the last twelve months to code that's literally been built twenty-five years ago. And so, that frequency of change depends on the code level and the solution that's running in the store.

We're trying to get it down to in the order of updates every month when it comes to kind of core software packages. And down to literally every ten, fifteen, twenty minutes when it comes to alerts coming off of the devices and us taking automated remote actions on those. It's a wide spectrum. And that's just in the store, and then we've got to think about 'above the store'.

CUOMO: Yes, tell us more.

VANDERHAM: We started thinking about online ordering. When I order online, it sends in the solution, somebody goes and picks it—"Oh, the steak that I want is out today." You have to be able to real-time interact with the consumer, again, through an automated way, using the picker device in the store and me through my texting or chat app that I'm using on my mobile device. And so, we're automating that. The above-store stuff we are literally updating multiple times a week. You see this paradigm of a hybrid environment. We've been talking about hybrid in software for a many of years.

CUOMO: Sure, sure.

VANDERHAM: This hybrid environment, what runs in the cloud, what runs in the store and then how we manage those, I'll say homogeneously, but on different intervals because of still some of the complexity that comes. And then maybe during the session, we can talk about some of the things we're doing at the edge to made edge look more like cloud in the long run.

RETAIL ON THE CUTTING "EDGE"

CUOMO: Let's talk about edge. What's the role of edge here, and what's running at the edge? What's the interaction between the edge and "the mothership," and how does automation play a role in that?

VANDERHAM: Yes, about three years ago now, we acquired a company called Zynstra, where we started virtualizing at the edge. We could take those old Microsoft Windows monolithic code bases and at least virtualize them so that you can shrink-wrap it in a VM (virtual machine), test it, configure it, and deploy that and do it in almost real-time—no downtime—updates. So, we're running a lot of that at large retailers today.

One of our great examples is Pilot Flying J. They have over five hundred locations for truck stops—everything in their store runs virtualized in the back of the house with a dumb thin client on the front end. They're on the leading edge of this virtualization. That's what we've been doing for the last three years. And now the team's working on how to break apart those monoliths into microservices, how do we run that application or that set of services, and how to deploy a set of components at the edge in a container-based model?

We're using edge to replicate MicroK8s infrastructure and running sets of microservices. So, think of a selling engine, think of a tax engine, think of an inventory engine, and then all those run-in store in the back of the servers with a lightweight, thin client. And then we're going to be able to update those in not a monthly cadence but in a daily cadence and manage SLAs at the edge.

AI IN THE RETAIL WORLD

CUOMO: Hey, Tim, let's jump to tracks here. This is *The Art of Automation*, where we talk a lot about the relationship of AI and automation. Could you add the AI perspective to the story here? What's the role of AI in automating retail?

VANDERHAM: AI serves a huge role because we have to be able to take all of the sensor data off of the hardware, the software, even human interactions. When you start thinking about how we can instrument our mobile devices or our websites, we can start to see people's activities, right? And leveraging that super set of data — again, machine data, software data, transactional data—which is also really important—and then consumer behavior data.

Bringing all that data back to a common data lake, leveraging AI and advanced machine learning algorithms against that to make sure that we are doing the right things for our customers. NCR is committing to our customers an SLA— a service-level agreement of when your systems are up and running, able to transact, and that you're able to transact twenty-four hours a day, seven days a week in many cases.

We have to use AI to intelligently filter out the noise so we can focus on what really needs to be fixed, because let's be honest, there are so many of these machines. Sometimes we have to literally filter out the noise, get the telemetry data back, and say, we've got three self-checkouts down out of six at this one store for various reasons. Get a customer engineer in a truck rolling to that store to get them back up and online for when they open at 6:00 a.m.

So, that's the level of sophistication that we're working through of hundreds of thousands of events every day from every retailer, millions across our retail set. How do we filter noise out, how do we give support and, ultimately, how do we deliver availability? So Jerry, when you're a consumer and you want to walk into Whole Foods or you want to get your gas at Sheets there in North Carolina—both good customers of ours—you're able to transact because our systems have made sure we're available.

CUOMO: I've seen a situation where I was about to scan a bottle of wine, and it could have been my imagination, but it seemed like the checkout manager got an alert to ask for my license before I even pulled it out of my cart.

Are we in a world where there's that level of granularity of awareness in the store? Is AI—things like computer vision—is it really an alternate employee now in the store? Maybe paint a picture of where this is going. What's a day in the life of a consumer going forward?

VANDERHAM: Yes, you're absolutely right. We call your self-checkout environments "bullpens." When you think about all the cameras in the bullpen—yes, there's a camera in the scanning agent in the self-checkout machine. Many of them have a camera above head now as well to avoid shrink, and that could be what's picking up objects that are in your basket.

So, this isn't widespread yet, but when you start thinking about that, yes, the eyes in the sky around those bullpens are going to help us understand when baskets are really empty and when they're not so they can avoid shrink. They're going to hopefully help identify you, Jerry, as who you are, and if you're a loyal member.

Some of our retailers are wanting facial recognition so that they know that "hey, Jerry, you're back in today, thanks for your business. Oh, by the way, we're going to give you a special discount on something that will show up in your mobile device as well just because you came in today."

Heatmap, tracking, natural language understanding are all at work – but at the end of the day, we all want a better consumer experience. So, we're going to leverage computer vision, facial recognition, advanced AI concepts if you opt in—obviously, from a data security perspective—and if you opt in, we will give you a better experience, which also makes that consumer better able to serve you day in and day out.

CRYPTO IN RETAIL

CUOMO: On that topic of painting a picture of the day in the life of a customer, you've been in the news recently for your acquisition of LibertyX, and I think being a very progressive point-of-sales organization thinking about crypto, can you quickly comment on some of your ideas around crypto and what you're thinking?

VANDERHAM: Yes, absolutely, Jerry. So, you're right, we acquired a company called LibertyX—they're a crypto-based company for three things. We can allow consumers to buy and sell crypto in a digital format or with cash at a physical device like a self-checkout machine or a point of sale. We can help consumers using crypto as the rails, which is a more consumer-friendly way and a safer way to move money from, say, to the U.S. to Brazil or U.S. to Mexico.

Finally, it's payments. I want to allow consumers to be able to take whether it be Bitcoin or other alt coins or a stable coin like USDC, and whatever you have in your wallet, if a consumer wants to interact that way, we have to enable that at that point of sale at that self-checkout machine.

That's our vision, and I like to say we want to put crypto on every corner, because we have hundreds of millions of touchpoints every day digitally and physically with consumers. So, Jerry, next time you want to buy your groceries or your gas with Bitcoin or other cryptos in the future, look for an NCR pump or an NCR point of sale, and you'll be able to do it.

CUOMO: Will do, Tim. Thank you so much, that has been inspiring, impressive, and so progressive. Thank you so much for spending time with us this afternoon.

VANDERHAM: Thanks, Jerry. Great to be here, and thanks for the time. Always great to see you.

UP NEXT

As Tim stated, "automating retail is really out of necessity".

Retail is under pressure. Margins are stressed from all sides: higher costs to manage e-commerce supply chains, growing demands from suppliers to pass on raw-material cost inflation, higher investments to match new competition, and steadily rising labor costs. At the same time, the customer's expectations continue to surge as digital natives and disruptors alike raise the bar for personalized service—on the back of what, at times, is an advantaged cost structure. As retailers struggle to adapt, and even to survive, they increasingly pursue automation to address margin strain and more demanding customer expectations. Automation will shape retail business models and the broader value chain, creating organizations with fewer layers and a better trained and trusted workforce empowered by real-time data and analytics. The winners in the sector will be those who understand these implications and act quickly to address them.

The above paragraph is from a McKinsey & Company article that is a great resource to learn more about automation in Retail.[38]

* * *

Make sure you also check out *The Art of Automation* podcast, especially Episode 21, from which this chapter came. Listen in to the full episode as Jerry and Tim they discuss how progressive retailers are beginning to use automation to transform customers' shopping experience.

Chapter 11 – Figure 1 – Podcast Episode 21 – Automation in retail

Chapter 12
FINANCIAL SERVICES

A conversation between Jerry Cuomo and Oscar Roque from Interac Corporation on automation in the financial services industry.

COVERED IN THIS CHAPTER

- Automation in the financial services world

- Ingesting a world of data

- Giving back the gift of time

- Financial services as an industry crossroads

AN INDUSTRY LENS

This chapter of *The Art of Automation* is a reduced transcript of a conversation between Jerry Cuomo and Oscar Roque, vice president of Strategy & Emerging Solutions at Interac Corporation. In our final chapter of AI-powered automation viewed through the lens of a specific industry, Oscar and Jerry discuss automation in financial services from the ATM to AI.

JERRY AND OSCAR

CUOMO: Welcome to *The Art of Automation*, a podcast that explores the application of automation in the enterprise. Today we're set to explore automation in the financial services industry; and folks, AI-powered automation is already well at work there. It's transforming how these organizations operate internally and enhance interactions with their customers, like me.

And I don't know about you, but I often use my bank's chatbot to check my balance, make payments, all sorts of things. I just love it. And think about it: there are iconic automation systems that have come from this sector, most notably, the ATM, as in "Automated Teller Machine." There it is, right there in the name. And ATMs came in existence in the 1960s.

Fast forward to today. Artificial Intelligence is now in the mix, and it's attracting huge interest as the financial sector explores its potential to unlock value, to improve customer satisfaction and loyalty, more responsive and effective operations, and tighter compliance and fraud detection.

For today's episode, I'd like to invite an industry colleague of mine that I've had the pleasure of sharing a stage with before: Oscar Roque. Oscar is the vice president of Strategy and Emerging Solutions at Interac Corporation in Canada, with a top-line mission to "shape the future of finance." Oscar has led projects in open banking, digital payments, blockchain, and fintech. Welcome, Oscar, to *The Art of Automation*.

ROQUE: Thank you very much, Jerry. It's very good to be here and to speak to you once again.

AUTOMATION IN THE FINANCIAL SERVICES WORLD

CUOMO: Let's get right to our first question if you don't mind. Oscar, why are you so excited about the opportunity to shape the future of finance;

and, why now? How does what's happening in the world affect "the shape," so to speak?

ROQUE: Yes. It's a great question. When I think about finance, I think finance in its current definition has change, especially with the pandemic, it's no longer about just finance. We've already seen quite a significant shift to digital in everything that we do, whether it's finance, the day-to-day interactions. It's just all around us.

And when I think about digital, you would normally classify it as a certain group or certain segments of folks transacting digitally. But the benefit of the pandemic is that it really transformed what finance is, what other sectors are, and it has really started intertwining them together.

And what this really means across the consumer, and the end user, into small businesses is the ability to start blending those experiences into "what's right for that consumer." We're a payments company, naturally, so as an example we're foraying into digital ID. And when we think about the end consumer and their experience and their journey, we make sure they're at the very center.

And so what a digital shift means, it's more than just payments; it's how they transact digitally, how they log in and how they show themselves, how they verify themselves, and all those important pieces.

And it's really exciting just to be in finance because it doesn't mean to be limited within financial services anymore. And I think that's just awesome for the consumer at the end of the day.

CUOMO: Yes, I agree. And when you say "digital," I assume you mean, multimodal—customers coming in through mobile apps and everything in between? Could you maybe share a little bit about that?

ROQUE: Absolutely. It's the blend. There's no more physical, there's no more different types of digital; it now all rotates around the consumer, and whenever and however they decide to transact with whatever they're trying to do, whether it's a person or even a device.

And we're seeing this, sure, in the financial services world, but we're also seeing this relevant to how we're conducting even just meetings. We are so used to face-to-face, but just given the circumstances, we had to quickly adapt and shift, obviously, to virtual meetings.

But then you transpose that and you think about it, you realize conferences have fundamentally changed as well. Whether they're hybrid or fully virtual, the way you conduct panels, the way you have to change your behaviors and your body language that it really translates digitally and in one and zeros versus what you didn't have to do before the pandemic.

This is a paradigm shift right across the board. And that's pretty exciting just in terms of what it means for digital.

CUOMO: And at this point, there's no going back. I'd like to also say that being digital is the precursor to automation.

ROQUE: Hundred percent.

CUOMO: It opens the door. Once you're programmable, now you invite the creation of new programs. And one of the more interesting types of programs are automation programs. Can you tell us a little bit about automation and financial services?

ROQUE: I love this piece that you mentioned around being digital is the entry point to automation. This is because that digital transaction, no matter what you're doing, whether it's a meeting online or whether you're transacting with something or someone, there is data now that's being generated out of it. And from that data, yes, it absolutely leads to the ability

171

to automate and just make things more seamless and drive a better consumer experience.

I've got the fortune to be able to lead corporate strategy for my organization, and so when we start to think about that, we realize wow, there's a lot of data out there. And as Canada's trusted payments company, how can we make sure that we help Canadians at the end of the day have full transparency in how data's flowing and also help them to think about control and consent of that data? Ultimately what could happen is with the proper setup of that data flow, the proper benefits can be driven through automation.

INGESTING A WORLD OF DATA

CUOMO: At the top of the podcast, I mentioned how I was excited about using my bank's new chatbot as a way to make payments and things like that, but I think being digital is more than just about chatbots. Can you talk a little bit about capitalizing on the digital investment, perhaps using technology like AI, and how do you really take your customers' experience to the next level?

ROQUE: Yes, it's a very interesting piece. I tend to think of it with a two-pronged approach. There's definitely the normal everyday conversation that most of us corporate folks have around the table regarding how you can actually help the customer with this increased data flow—obviously, with their consent—and how we can derive more value for them. And how we can make those experiences seamless, whether it's dealing with how you manage your money, whether it's making sure that the fraud is all preempted and you're catching it ahead of time, and you're not disrupting the experience—wherever they are. And the list goes on.

But I also think about a second angle, which is around data flow and automation. And so, since my team is leading strategy for our organization, what's fundamentally important is a lot of research and a lot of inputs to drive strategy. Such competitive intelligence requires a significant amount of ingestion of data or articles. Then it's about the digestion of it, and then, synthesis of that data back in a very succinct way that is relevant to the business.

When I think about automation, the more complex the world gets, the more these sectors are all convening or becoming one, and it's no longer about just financial services. What that just means for a strategy team or a data team or a research team is that there's a lot more to ingest because something in the healthcare or the telco world could definitely now significantly and directly impact in the financial services world.

And that's probably stating the obvious, but that just means we have to be able to get ahead of it as a strategy team and to figure out, we can't just add bodies to this, we need to be able to automate it, to your point.

CUOMO: That's right, and you've made a big impression here with a different variety of data sources. I would say some of that data is structured, some of it is probably noisy, unstructured data. So you have to kind of connect the dots, and, you know, all the analogies, finding needles in the haystacks and all of that to make sense of it.

Tell us a little bit about how then perhaps this is where machine learning and other technologies come in to help make order from that chaos of data.

ROQUE: It's funny, when we think about data generally there's definitely the two camps: there's structured and unstructured. But when you end up ingesting it into your organization, if it's not all structured the same, it just looks like it's unstructured—it feels like it's unstructured. It's basically an exercise of restructuring everything regardless of its structure outside.

And so we end up looking at data really from three different categories. One such category is what we call the forefront or the academic type of category. These are things like scientific papers, to scholarly papers driven out of universities, colleges, et cetera, patent offices, for example, and even conference presentations.

The second category is what we call "journalistic." This is what you read in the newspaper, this is what you see online, all those news sources. And again, they may be credible, or they may not be, but this is also part of the learning language, if you will.

And then the third category is social media. This is the most unstructured data, but again, when we treat it as it comes into the organization, we treat it all the same across the three categories.

And so the interesting piece about all of this is how do we take these three things which are only exploding and expanding and be able to say, "Oh my, how do we create competitive intelligence and make sure that it's right and it's relevant?" And this is where machine learning comes in.

And as you could imagine, we have a bench of several researchers who are doing this day in day out for a living, forty hours a week, fifty-two weeks a year. Where I see the opportunity for AI, at the very least, is to be able to train—let's call them "AI agents"—to augment and add to the team, and then, maybe one day be able to replace some of those team members. But not literally replace them, but be able to have those team members use their brainpower and knowledge capital and apply it to higher order things that require human judgement. So, the AI agent is their proxy, automatically ingesting, digesting, and synthesizing while they sleep. And that is really exciting.

GIVING BACK THE GIFT OF TIME

CUOMO: You're right Oscar. It's about building the hybrid work-force. Technology and humans doing something better than either of them individually. That's great!

With that, can you paint a picture of where this is all going, maybe even a day in the life of one of your employees or customers. How it might all come together?

ROQUE: A day in the life. I just absolutely love this question. It's a great way to be able to make things real and it's a great way frankly to look at the applicability of how automation and AI can actually impact us every day.

Imagine a world where there is so much information and misinformation… and perhaps this world is much like today, but much more exacerbated. With an endless set of [data] sources… whether they're legitimate or not. But as humans we have limited time and brain capacity to process all this naturally. So in comes this ability to automate research [which is the job my financial services strategy team at Interac is responsible for] or more precisely use an "AI agent" to be able to help synthesize all of [the worlds] latest information and news in a manner that you would if you could actually pause time, however long you need [to ingest and digest that vast stream of information].

As an example, whether you're asleep or you're busy with other work-home tasks, this AI agent could easily search and scour the web, [or even other sources] and then [after netting out the salient information in an orderly way], it could update you with all the pertinent and relevant information in the news that you missed while you were focused on that other task.

Imagine this model applied to an entire team [again, like my strategy team] responsible for financial services research. Imagine how this [significantly enhanced and vast] insight can be meaningfully applied to you and your organization… and then to your customers.

Such and AI agent wouldn't just be limited to traditional sources. It would naturally expand to non-traditional sources, [especially useful when those sources are unstructured], like Reddit or other social media platforms. And not only checking for the information but also the velocity of information.

Imagine being able to detect GameStop [who experienced a social-media-fueled increase in share price] much earlier because of the velocity of [social-media] posts. Imagine being able to learn about investing potentially before the end of the year, rather than when [the stock] blew up earlier the following year.

Amazing! And what does this ultimately mean? Well, this AI agent can free up our time so that we focused on other matters… more important matters… [the "macro" matters]. Think about the impact when we can spend more time on matters related to equity diversity inclusion (EDI) and/or environmental social and governance (ESG). Specifically, the things that require a lot of discussion. And the things that are important for the future of our society as a whole. And how solutions might positively impact our kids and our kids' kids.

This is what actually really excites me about the "power and the art" of AI and automation.

CUOMO: Oscar! Wow! Drop the mic. Not sure day-in-a-life can be painted again so perfectly. And I love that idea of an AI agent doing research work for you when you are busy and doing other things.

But if I can just squeeze in one more question?

FINANCIAL SERVICES AS AN INDUSTRY CROSSROADS

CUOMO: Oscar, you and I have worked together before, and I'd like to reflect on some of the projects where you're acting as "a joiner" of industries. I don't know if you can talk about some of that, doing things around sustainability, working with the utility companies, and bringing together payments in interesting ways. Maybe you can reflect on how your industry acts as a facilitator or crossroads of sorts. Because once you automate your enterprise, I think you also have the possibility to create networks across other enterprises.

ROQUE: Yes, a hundred percent. And most people will know Interac as Canada's domestic debit network, and we run a variety of different platforms that allow Canadians at the end of day to be able to spend their money wherever, however, and whenever they want.

And the interesting piece about this whole network effect that you were talking about is we don't only see the transactions as being a payment transaction. A transaction can evolve to being a value transaction across our network, across our reach, and so that certainly applies to topics like digital ID.

But also, one could say that also applies to anything that's data related. When you talked about some of these earlier projects that you and I worked on, it's, we worked with a utility company, we worked with a healthcare company. And at the end of the day, we were able to use our network to be able to provide value, so that when consumers shared their data and did good for themselves or the planet, they actually got value and return for it.

And that's great when you think about it because there's an element around incentives and behavioral nudging, but there's also a higher order element around how do the actions of one or the actions of many from a day-to-day perspective end up evolving to be a significant community movement. And that's pretty exciting.

CUOMO: I agree. So, ladies and gentlemen, you've been listening to Oscar Roque, the innovator, the shaper of the future of finance, here at *The Art of Automation*. Oscar, I want to thank you for joining us today.

ROQUE: Thanks, Jerry. Glad to be here.

UP NEXT

After listening to Episode 23 of *The Art of Automation* podcast on this very topic, a listener (yes, it was the same listener who asked the question on insurance) asked if Jerry could provide a comprehensive list of "the sorts of things" that could be automated in a financial services industry context. Once again, working with colleagues from IBM Consulting, an extensive list was produced, and we thought it would be interesting to share that list here. Like before, the listing also attempts to estimate the "typical labor savings" for each category in the list. The following figure is a snapshot of that list, which can be summarized by the statement "just about everything and anything can be automated."

CUSTOMER ORIGIN | VERIFICATION	CUSTOMER INTERACTIONS
• Rates and Terms	• Customer Service
• Application data collection	• Customer on-boarding
• AML / Sanctions Monitoring	• Product on-boarding
• Risk Review	• Product cancelation
• Credit Evaluation	• Customer account closure
• **Typical Labor Savings: 25%**	• **Typical Labor Savings: 25%**
BUSINESS ADMINISTRATION	**CASH FLOW MANAGEMENT**
• Procurement/Vendor Management	• Reconcile Cash Transfers
• Business Process Management	• Billing / Collection
• Asset Management	• Payments / Funds Settlement
• Technology Management	• Loans & Dividends
• Auditing/Legal/Regulatory	• Financial Reporting
• **Typical Labor Savings: 25%**	• **Typical Labor Savings: 25%**
COMMERCIAL | TRADE FINANCE	**RETAIL BANKING**
• Working Capital	• Auto Loan
• Lending against raw material	• Mortgage credit decisions
• Line of credit/guaranty	• Credit/Debit cards
• Lending against finished goods	• Loan against securities
• Invoice discounting	
• Supply chain financing	
BUSINESS ACQUISITION	**PRODUCT MANAGEMENT**
• Campaign Management	• Market Research & Analytics
• Campaign Execution	• Product Deployment
• Targeted Offering	• Promotion and Brand Management
• Content Delivery	
• **Typical Labor Savings: 40%**	• **Typical Labor Savings: 10%**

Chapter 12 – Figure 1 – List of automatable tasks for financial services

Make sure you also check out *The Art of Automation* podcast, especially Episode 23, from which this chapter came.

Chapter 12 – Figure 2 – Podcast Episode 23 – Automation in financial services

Chapter 13

AUTOMATION AND
THE WEATHER

A conversation between Jerry Cuomo and Lisa Seacat DeLuca, distinguished engineer, author, and one of the most prolific inventors in the history of IBM (with over eight hundred patents) on automation and the weather.

Interview by: Jerry Cuomo

COVERED IN THIS CHAPTER

- Technology in our environment

- No energy wasted

- The weather's impact on business

- The five-year forecast

WHAT'S THE FORECAST?

This chapter of *The Art of Automation* is a reduced transcript of a conversation between Jerry Cuomo and Lisa Seacat DeLuca, IBM distinguished engineer and director of emerging solutions for Weather & Agile

Accelerator. In this discussion on one of the most unique applications of AI-powered automation, Lisa and Jerry discuss how businesses can use automation to understand the environment and use that data to tackle challenges related to climate change, sustainability, and everyday business operations. They also make predictions for how automation will play an even bigger role in public health as the world transitions to the next normal, with technologies coming together to help us be safer as a society.

JERRY AND LISA

CUOMO: Welcome to *The Art of Automation*, a podcast that explores the application of automation in the enterprise. Well, this is a big day for *The Art of Automation* because I'm delighted and privileged to discuss automation in the context of weather and the environment, with Lisa Seacat DeLuca. Lisa is currently focused on modernizing our weather business solutions in the aviation portfolio. (Yes, IBM has aviation solutions!)

Lisa is a rockstar. She's a TED speaker and a self-published author of two children's books, titled *A Robot Story* and *The Internet of Mysterious Things*. Lisa's one of the most prolific inventors in IBM's history. Her invention portfolio includes over 800 patents filed, of which 550 have been granted to date. What an honor this is! Welcome, Lisa, to *The Art of Automation*.

DELUCA: Thanks for having me. I'm excited to be here.

CUOMO: So, let's jump right in—I've got a lot to talk to you about today.

Lisa, your current work is focused on innovation and, specifically, applying technology to understand and improve the environment. Tell us why you're so excited about that.

DELUCA: The environment is exciting because it really affects all of us and every single business. Everybody's affected by weather, and some say that the climate and how the future is going to look around the environment. And of course, hot topics in the news right now—sustainability and climate change. It's just an exciting place to work in.

TECHNOLOGY IN OUR ENVIRONMENT

CUOMO: Let's talk about the technology. This is *The Art of Automation*, so I'm going to start asking you questions about how automation plays a role, how artificial intelligence (AI) plays a role. So, let's talk a little bit about the technology behind the scenes here, Lisa.

DELUCA: Yes, yes. To me, when I think of automation, I think about incremental simplification, right? You think of like a drive-through, and that's just automating or making it easy for you to get food for hungry customers. And then you can further automate it by other solutions like Uber Eats or DoorDash or Postmates. Basically, you're just making it more convenient. How do you get food to a hungry person?

As an inventor, when I think about automation, it's really just this innovation, that incremental improvement to processes. When we think about environment and what we're doing with weather, it's just so data rich, right? We have data coming from all over the place—historical weather data, future weather data—and it's trying to take all that information and overlay it with other factors that might be outside of weather to try to come up with business impacts from it.

CUOMO: Lisa, I just have to ask you: Where does the weather data come from?

DELUCA: The answer is from all over the place, right? Some people have personal weather sensors, there's external data, there's data coming in from the government, satellites—everywhere. There's so many different sources all coming together.

CUOMO: And your team is aggregating this data in some shape or form?

DELUCA: You got it, yes. It's a lot. We've got terabytes of data on our geo-spatial analytics product, right, so we can layer things like COVID data, flu outbreaks, mosquito data. All of that could be overlaid on top of other geospatial analytics.

CUOMO: All in the cloud, I would assume?

DELUCA: All in the cloud, yes. Our solutions are all SaaS-based, so it's really fun to see how quickly you can compute and make business decisions from them.

CUOMO: And Lisa, this is The Weather Company behind this?

DELUCA: You got it. The Weather Company. So, internal in IBM, we're split to the consumer and business sides, and I run our business side. And we've got industry solutions, so oil and gas, energy and utilities, think of railways and tunnels. It's so cool how it does really impact every business.

NO ENERGY WASTED

CUOMO: So, automation starts with data, and we have sensors all around the world feeding the weather cloud. And then how do we apply technology, AI, etc., to make a difference?

DELUCA: Sure, I'll give you an example. One of our solutions is around vegetation management, which is kind of fun, because growing up you see trees everywhere, you see the power lines, and you never expect so much technology to go into it. But we use satellite imagery to understand how tall trees are. So, we can predict the height of trees and then you can help prioritize where you need to trim trees, open up work orders for people to go out and trim the trees.

CUOMO: So this way, you know their trip out there is going to be for a purpose.

DELUCA: Exactly. And throw in some weather data, it's like ah, don't go right now, there's a storm happening. Let's do it next week.

CUOMO: Yes. Automation is reducing the repetitive, mundane work so us humans can focus on things that matter and have a better outcome for the business, right? And I think that was a pretty cool example of that. You're using data. You know how tall the tree is and voilà, no energy wasted—no pun intended.

DELUCA: Exactly.

CUOMO: What about artificial intelligence? What role does it play in the environment and weather?

DELUCA: For us it's really the machine learning side of it and just the sheer amount of data. All the prediction that you're doing with weather, right? Is it going to be 50 degrees tomorrow? Is it going to rain? All of that is all AI and machine learning based on all the historical information we know about what happened in the past. And throw in climate change, and it's really hard to predict, so you need even more data, even more factors to be accurate.

THE WEATHER'S IMPACT ON BUSINESS

CUOMO: You mentioned predicting the weather. And every morning we all sit through some level of that, so thank you for helping with our morning forecasts. But what else could we do? What other hints can this weather analytics bring for business? Can you jump the tracks from weather for the sake of weather and weather for the sake of business and automation and AI?

DELUCA: We're starting to look into things like carbon accounting, which is super important when you think of the supply chain use cases. So, using that carbon footprint, understanding the kind of products that you're putting on either a truck or a plane or a train and how that's going to affect your business.

And also climate risks, so we're pulling in things like wildfire risks, as well as flood risk. Again, that helps you decide what to do with your business, whether or not to prioritize maintenance on one stretch of highway or another stretch of highway, all based on those other factors.

CUOMO: And how would this work if you were an insurance company? You would subscribe to a stream of data and analytics? How would someone participate in that?

DELUCA: Exactly. Some really cool new technology that I've just been playing with is LIDAR data. Rather than just the satellite imagery, the LIDAR data kind of breaks it up and almost creates a 3D effect. So, imagine the insurance use case like you brought up, you can see it down to the roof what the damage was before a storm came in and then afterwards, so that you can accurately pay out claims and understand what the damage was of that event. All of that from the technology. I just think it's so cool how something that seems so unrelated to different industries can make such a big difference.

CUOMO: Lisa, I also know you've done work with real estate planning, and I'm just thinking about back to work and environmental COVID and coming back to our next normal. What are your thoughts on applying technology, automation, AI—what are the challenges and what are the opportunities?

DELUCA: Yes, there are so many opportunities and challenges. You've seen how people are trying to do vaccine pass to show who has already got the vaccine. I think we're going to a lot of that, right? Show proof that it's safe for you to go into the office. I've gone into buildings where they're doing a screen on my temperature and making sure you're all right in that regard. There's contact tracing.

It's really exciting to see how all the different technologies are coming together to help us be safer as a society and know that it's okay for you to come in. I think we're going to see a lot of people going in the office and a lot of people are enjoying the simpler life of being at home, as well. It will be interesting to see how offices are repurposed and how people come together for conferences and how the technology is used to help make you feel safe.

THE FIVE-YEAR FORECAST

CUOMO: Where do you see the world going in two, three, five years? You set the timeline, but how do technology, insights, AI impact our future? Paint a picture for us.

DELUCA: I think AI is here already; it's here to stay. It's not going anywhere. People are going to probably stop talking about it because it's just going to be normal; it's just going to be you're used to AI being applied to all of your business decisions, so it's like what's the next thing after AI?

And automation is here to stay, right? That simplification, the incremental improvements in innovation. Back to your input about automation, it's just giving us more time—more time to do other things than work, more time to spend time with our families. I think that's what the future is going to be.

CUOMO: Lisa, you once shared an example with me that I'd like you to share with the audience about flight arrangements and automating flight arrangements using telemetry from weather, etc. Can you share that?

DELUCA: Yes. Today a lot of airlines use paper—

CUOMO: Wait, what year is this?

DELUCA: I know, I know. Can you believe it? And I remember being at the airport and hearing the printers going, and it was even the old printer machines that had the holes on the side.

But they get that piece of paper and they give it to the pilot, and then the pilot has to sign off on everything about the aircraft, right? Is it safe to operate? And so we're working on a solution around a digital flight release, which is really exciting because you can imagine you can take sensors from the airplane itself and understand fuel levels and understand how much it's operated, what that flight path is going to look like, who's on board, how much are all the people going to weigh? What are our carbon emissions? So many different factors you can pull in instantly for that pilot to do their digital flight release.

CUOMO: And is there a time savings here?

DELUCA: Definitely. Every minute saved for airlines is like thousands of dollars. So definitely it's amazing how automation and that time saving really does translate to money.

CUOMO: Lisa, if I can, you said the environment is a hot topic these days—no pun intended—where are we going from here? Can you share your personal views on technology and what we can do—the art of the possible for the environment?

DELUCA: I know, sometimes thinking about the environment you feel like, you're just one little person; how can I actually contribute or help save the planet? But I think the first step is making companies accountable for how they're contributing and how they're impacting the environment. There are now reports that go out and companies are proud to show that they are getting towards net zero or their carbon emissions are low. So, you're going to see more of that, more pressure that the companies and other companies are putting on themselves that will lead to better output for the world.

CUOMO: Well, thank you so much, Lisa. That was both inspiring and insightful. Thank you for joining us here at *The Art of Automation*.

DELUCA: Thanks, Jerry. Thanks for having me.

UP NEXT

Lisa clearly has a gift of creativity, as heard through this interview and witnessed by her hundreds of patents. Therefore, we would be remiss if we didn't take a closer look at one of her patents related to AI-powered automation. Here is a quick look at one such patent related to geofencing, which applied to the vegetation management and LIDAR scenarios that Lisa described above to automate the dispatching of workers to clean up after fallen trees (or powerlines, or…).

Cognitive geofence updates[39]
Patent number: 10785598

Abstract: Methods, computer program products, and systems are presented. The method computer program products, and systems can include, for instance: examining data of breaches of a geofence by client computer devices to determine respective positions of the breaches; establishing an updated location for the geofence using the determined respective positions of the breaches; updating a location of the geofence so that the location of the geofence is the updated location; obtaining data of a client computer breach of the geofence at the updated location; and providing one or more output in response to the obtaining data of a client computer breach of the geofence at the updated location.

Jerry Cuomo, et al

Type: Grant
Filed: February 28, 2020
Date of Patent: September 22, 2020
Assignee: INTERNATIONAL BUSINESS MACHINES CORPORATION
Inventors: Lisa Seacat DeLuca, Jeremy A. Greenberger

* * *

And make sure you check out *The Art of Automation* podcast, especially Episode 14, from which this chapter came.

Staying on the topic of automation for the environment, we have one more use case to explore, which is the use of AI-powered automation to navigate an autonomous vessel across the ocean. Like the previous topic, this is an exciting look of how automation can be used to do things only once seen in science fiction movies. As you read on, we suggest that you also think about how the ideas surrounding automation of the environment might also be applied to the enterprise, which is a topic covered in the next interview.

Chapter 13 – Figure 1 – Podcast Episode 14 – Automation and the weather

Chapter 14

AUTOMATION AT SEA

A conversation between Jerry Cuomo and Don Scott, director of engineering at Submergence Group and the mastermind behind the *Mayflower Autonomous Ship*, on automation at sea.

Interview by: Jerry Cuomo

COVERED IN THIS CHAPTER

- The coolest job in tech

- The "AI Captain"

- Rules and regulations at sea

- From the ocean to the enterprise

FULL STEAM AHEAD

This chapter of *The Art of Automation* is a reduced transcript of a conversation between Jerry Cuomo and Don Scott, director of engineering at Submergence Group and the mastermind behind the *Mayflower*

Autonomous Ship.[40] In this special discussion on perhaps the most famous IBM engagement since *Jeopardy!*, Don and Jerry discuss the role AI-powered automation is playing in the *Mayflower Autonomous Ship* project, which is attempting to become the first entirely automated vessel to traverse the Atlantic Ocean. They elaborate on the "AI Captain," safety, trust, and why AI explainability is so important. Jerry and Don close by exploring how lessons learned from automation at sea can be applied back to an enterprise setting.

JERRY AND DON

CUOMO: Welcome to *The Art of Automation*, a podcast that explores the application of automation in the enterprise. Today, we have an exciting topic that, on its surface, may seem a bit out of the boundaries of classic enterprise. This episode will explore a unique application of automation, which is to traverse the ocean in a quest for data and discovery. As you hear the story unfold, I'm sure you will identify with as many similarities as differences between automation at sea and automating your enterprise. Trust me, you'll see this.

And yes, I'm talking about the *Mayflower Autonomous Ship* (or, *MAS* for short). *MAS* is a grassroots initiative led by marine research nonprofit ProMare, with support from IBM. Their mission is to provide a flexible, cost-effective, and safe option for gathering critical data about the ocean. *MAS* can spend long durations at sea carrying scientific equipment and making its own decisions about how to optimize its routes and mission. With no human captain on board, *MAS* uses the power of, yes, AI-powered automation to traverse the ocean in a quest for data and discovery.

So, the ship's AI captain performs similar roles to a human captain—assimilating data from a number of sources; constantly assessing its route, status, and mission; and making decisions about what to do next. Cameras and

computer vision system scan the horizon for hazards and streams of mete-
orological data reveal potentially dangerous storms. Machine learning and
automation software ensure that decisions are safe and in line with collisions regulations.

For today's episode, we have the good fortune to have Don Scott with us.
Don is the director of engineering at Submergence Group, and we'll discuss automation in the context of this masterpiece of modern automation
technology, the AI Captain. Welcome, Don, to *The Art of Automation*.

SCOTT: It's great to see you again, Jerry. Happy to be here.

THE COOLEST JOB IN TECH

CUOMO: All right, let's just jump into the first question, if you don't mind.
Don, as I alluded to in the beginning, you have a pretty unique job. Can
you broadly share what you do at ProMare and the Submergence Group?

SCOTT: Oh, sure. I've worked my entire career as an ocean engineer. My
focus has been in the navigation and control systems for manned and
unmanned submarines for the U.S. and the U.K. navies. But for the last
fifteen years, my team and I have been developing marine systems with
different levels of autonomy, so, being able to operate unmanned.

CUOMO: Don, first of all, I think everyone out there is saying, "How cool
is that!"

Is this software, hardware, all of the above? That plus more? Can you tell us
a little bit about the tools of your trade?

SCOTT: ProMare and Marine AI and Submergence Group, kind of do
everything. My focus is on the software aspect. It's on the sort of intelligent

decision-making about where the boat should go and how it should operate, things like that.

CUOMO: Could you share a little bit of your favorites from a software tool perspective? Are you a C++ or Java person? What's your forte?

SCOTT: I started writing in ASSEMBLY. That's certainly not my favorite, I'll tell you that. We're actually a big sort of C# house, so we're kind of focused on in terms of our tools. The younger guys favor Python—not my thing—but that seems to be where everything's going.

THE "AI CAPTAIN"

CUOMO: Let's dive in. You said autonomous operations, and that's kind of like the highest level of achievement by an automated system. And your AI captain uses technology like computer vision, AI and rules, and edge computing all as a means to safely navigate around other ships, buoys, and other ocean hazards during this transatlantic voyage. Please share with us a little bit, if you can, about how you put these technologies into play to achieve this autonomous automation, and what did you learn while applying this technology?

SCOTT: The traditional role of a sea captain is the safe and efficient operation of a vessel at sea, and that's exactly the capability that we need to implement for true autonomous capability. *Mayflower*, or the AI captain on *Mayflower*, needs to evaluate the current situation and recommend a safe course and speed.

To do this, we implemented—call it a "hybrid AI solution." It's a classic design architecture for autonomous systems. We're not reinventing the wheel here. You have sensing, computing, and then acting. And what we've learned is the sensing is the real key, right? Like changing visual and physical information into something a computer algorithm can understand.

We're looking at video streams from six different cameras using computer vision capabilities; looking at AIS, which is a standard system for identifying ships at sea; radar, for looking around you; a fathometer for knowing water depth; and weather reporting; and importantly, chart information, too. We use all this information—we put this into what's called a data fusion layer where we, basically, pull all this information together to create what we call a hazard map. So, it just sort of defines the safe and unsafe regions at sea.

CUOMO: Very interesting, Don. And is this flexible? Does it learn over time? For example, I read something almost humorous a few weeks ago before the launch of the *Mayflower* that there were some fears about, uhoh, what happens if a seagull swoops in? Would the introduction of a seagull just kind of popping in and then popping out affect the algorithm. Is that something that you're planning for or you can add on the fly later?

SCOTT: Yes. I'm not sure where the seagull thing came in, but we're not too focused on identifying different types of birds. We were more focused on identifying different types of ships and what we call LOMOs—so, Low Observable Marine Objects— lobster pots, and basically flotsam that could be a hazard to navigation.

In terms of adaptability and learning, when we are doing the transit, we have a fixed set of inference models for the computer vision, but we've always retained the ability to push new models out to the ship as required.

We're not doing any retraining at sea. We don't want to be introducing potential new behavior during a transit, so we use a fixed set of inference models when we're operating at sea.

RULES AND REGULATIONS AT SEA

CUOMO: That makes sense, Don. And you had once asked me if I've ever talked to a marine regulator. I don't even think I answered because... that was the sort of question that I would never expect a person ever to ask me. No, I've never talked to a marine regulator.

SCOTT: I thought everybody talked to marine regulators—

CUOMO: [CHUCKLING] No, no. Sorry, not this person here.

But then you went on in a very interesting way to explain about international regulations for preventing collisions at sea and how our rules, language, and AI explainability, in general, establish a level of understanding, trust, increased confidence from these regulators. Can you share a little bit more detail for our audience here about how that works?

SCOTT: Yes. It's a pretty complex issue, to be honest. We need to start by understanding the environment that we're working in. Since we began sailing in the oceans, back when the Egyptians first had the first sailing boats with reeds, there was an understanding of how these ships would interact when they came across each other.

There's a fairly large organization called the IMO—the International Maritime Organization—which is a big international regulatory agency that's responsible for overseeing quite a bit about the ocean. And one of the things they look after is the safety of navigation at sea.

And they've created these conventions and rules and collision regulations, or COLREGs, that we like to call them. And like I said, these are the COLREGs, they dictate how ships interact at sea. It becomes a very complex situation because COLREGs only address single ship interactions—how two ships deal with each other—but anyone who's been to sea will know there's usually lots of ships around, especially in harbors and things like that. So, it becomes a bit of a dance, on how ships interact.

CUOMO: And is it fair to say that marine regulators can't read code— Python or C#?

SCOTT: Yes, absolutely. I'll certainly address that about how is it that we make these rules understandable to the regulator, and because the COLREGs, they can be somewhat ambiguous, right?

There's a favorite term of mine, something called "passed and clear," where when you're overtaking another vessel, you're not supposed to go back to your original course until you're "passed and clear of the previous vessel." What's "passed and clear"? It's when the other captain feels safe, right? So, when is that?

This brings up this whole issue of trusting the decision that an AI captain's making. And it's a pretty significant hurdle because they want to be able to understand not just the decision that's been made but also why the decision's been made, like the rationale behind it as well.

And that's the strength of using, for our solution, it's a deterministic rule set. We use something called ODM—IBM Operational Decision Manager.

CUOMO: Yes, I know it well.

SCOTT: And because of its pedigree in the financial services industry, it provides a completely transparent and immutable record of the decision-making process. And it's this transparency and explainability that are required to develop trust. This is a trust that's required for the regulatory boards to understand why we're making these decisions.

CUOMO: Do you literally show the regulators the rules, as written, actually makes sense to the regulators?

SCOTT: Well, we haven't had to yet, but we certainly will. We are certainly able to. That's kind of the power of this approach.

CUOMO: Wonderful. Yes, very interesting, very cool.

FROM THE OCEAN TO THE ENTERPRISE

CUOMO: Don, can you help us jump the tracks on how lessons learned from your experience building the AI captain to autonomous operations in an enterprise setting?

SCOTT: I guess first off, it probably would have been easier in an enterprise setting itself.

CUOMO: Yes.

[LAUGHTER]

SCOTT: But we decided to do this at sea. That has its own sort of difficulties or its own challenges, let's say, than working sort of just on a bench or whatever.

I guess one of the big things we did is we needed to containerize a lot of these different apps that we're running that make up the AI Captain. We have an expectation that systems are going to fail, and you can't have a critical system like COLREGs manager—which is what we call it—on a single point of failure; you need to be able to move it around.

And we needed to do that on the first transit attempt when we were in the recovery mode. We had a system failure, but because of the system architecture and the flexibility we were able to move these critical systems on to working systems and keep the ship operational and the way it worked.

The idea is, you can also push hot fixes up to the boat even while it's operational, which is a key element as well.

CUOMO: Yes, so resiliency in your enterprise, can learn a lot from resilience while making a transatlantic voyage. Very interesting.

SCOTT: One thing I would mention, though, is I think one of the real things that we learned, and I think this applies to not just the marine domain but also your enterprise domain, is that we needed to be able to manage the division between the expert knowledge of our master mariners and the programming of the rule set itself. You touched on that a little bit earlier.

We needed to be able to really effectively translate what at times seemed really ambiguous rules coming from the master mariner; and once they've been written into the rule set, be able to sort of mirror them back to the master mariner in a way they can understand, so they can say, yes, that's exactly what I meant, and understand that that's actually the rule that was going to get applied.

CUOMO: Well, Don, I have to say not only is this unique factor very high, its cool factor is off the charts, but also so is its applicability. If you squint your eye and tilt your head as you've been describing, it's quite applicable to a very broad set of industries. And I'm sure you and your team are quite proud of this accomplishment, and so are we. It's so impressive and inspiring. Thank you very much, Don, for joining us today.

SCOTT: Thanks, Jerry. I really enjoyed talking with you.

UP NEXT

In this interview, Don described the use of business rules as a key ingredient to producing the intelligence powering the AI Captain. There is a bit of a debate over whether rules are a more effective model versus machine learning for such applications. Rule-based systems and machine-learning models are widely utilized to make conclusions from data. Both of these approaches have advantages and disadvantages. It totally depends on the situation for which approach is appropriate for the development of business. Several business projects initiate with a rule or excerpt-based models to understand and explore the business. On the other hand, machine-learning systems are better for the long term, as they are more manageable to constant improvement and enhancement through algorithm and data preparation.[41]

Make sure you check out *The Art of Automation* podcast, especially Episode 18, from which this chapter came.

* * *

We are quickly coming to the end of the book, with only two chapters remaining. The next chapter is an interview that can be viewed as the "book conclusion" and "next steps" by providing a view of automation, past, present, and future. Read on....

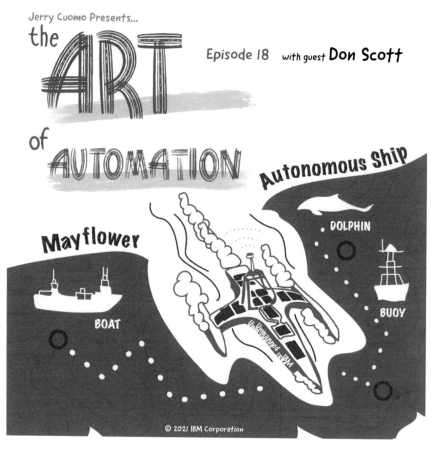

Chapter 14 – Figure 1 – Podcast Episode 18 – Automation at sea

Chapter 15

Yesterday, Today, and Tomorrow

A conversation between Jerry Cuomo and Ed Lynch, vice president of IBM Digital Business Automation, on the automation of yesterday, today, and tomorrow.

COVERED IN THIS CHAPTER

- Seizing today with automation

- Automation as the equal-opportunity employer

- Looking ahead

- It all starts with data

GAZE INTO THE CRYSTAL BALL

This chapter of *The Art of Automation* is a reduced transcript of a conversation between Jerry Cuomo and Ed Lynch, vice president of IBM Digital Business Automation. In this discussion, Jerry and Ed take a step back and discuss the concept of enterprise automation as a whole; specifically, where we are today and how we got here.

Ed explains that automation is everywhere—not limited to one specific industry. He then elaborates on the future of enterprise automation and how it inevitably will be centered around artificial intelligence (AI) and augmenting human work. They close by determining that automation all comes down to the human beings and that the best place to begin your automation journey is the data.

JERRY AND ED

CUOMO: Welcome to *The Art of Automation*, a podcast that explores the application of automation in the enterprise. So, what exactly is automation, you ask, and why dedicate a podcast to the topic? Well, in this episode, my guest and I are going to take a step back and reflect on this very question and do our best to shine some North Carolina sunshine on automation, looking at it in its recent past, where it is today, and where it's heading tomorrow.

So, if your business is dependent on being digital through and through—and I have yet to meet many businesses that aren't today—then listen in to the thoughts of one of our wisest wise men, Ed Lynch, who has been the visionary behind IBM's successful business automation strategy and product line since the early 2000s and is going strong today.

Few people that I've had the pleasure of working with balance the business and technology of automation the way Ed does. He has taught me so much about this space. In this episode, Ed is going to share some of that wisdom with you in a hope that you can use that insight to better prepare for the future ahead.

And while Ed will reflect on the past and present of automation, I promise his views on where automation is going is worth the listen, including how human beings and computers will continue to evolve to do what they're best at doing. And with that, I'd like to welcome Ed to *The Art of Automation*. Welcome, Ed.

LYNCH: Hey, Jerry, good talking to you again. Thanks for having me.

SEIZE TODAY

CUOMO: So good to have this conversation with you. So let's get right into it. Ed, before we talk about where enterprises are going with automation, can you share a little bit of where they are today and why are people excited about it, and perhaps a little bit about why now?

LYNCH: Sure. I've been in this space for a really long time. Let me give you a brief glimpse of history and then tell you where exactly we are today.

I started this maybe in 2001, approximately. And back then, we were talking about system-to-system integration. We called it business integration, and that was really the synchronization of one system of record with another system of record. It was a very IT thing — it was managed with no humans in the middle, but it was a way of doing what we called contact sync way back then. And that was all about making sure that one system will work with another system.

And then we added human beings and then we added decisions and then we added planning, we added monitoring, and that's kind of where we are today. We've got a whole bunch of different mechanisms for automating work—whether the work is human work, whether the work is system-to-system work, whether the work is dashboarding work—that's kind of where we are today.

We've got this trend in the marketplace which is really prevalent—and it's become extremely prevalent with COVID—which is the hyper-automation trend. Gartner has declared "hyper-automation" the number one trend now for two years in a row. The hyper-automation trend is why there are billions of dollars coming into the marketplace. Just yesterday, there was an announcement that the company Celonis had just secured a billion dollars' worth of funding.

CUOMO: Wow.

LYNCH: Like just completely, completely crazy. But why? Why now? Well, it's because everybody's focused right now on productivity. The productivity of the workers, not productivity of front-office workers with Office and Microsoft and that kind of thing, but productivity of the back-office workers, productivity of IT workers, operators, SREs, the system programmers who do work in automating IT systems. All of those different workers are under the microscope right now.

And businesses are looking for ways to get more out of the human beings and apply the human beings that are in both business and IT in really creative ways which take advantage of the skills that they have while letting the machines do the work that the machines are good at. Things that are very repetitive, right?

So, this whole space right now is going completely crazy. There are more unicorns in this space out in the market than you can believe, and it's all driven because of productivity. As businesses look at where they're spending human capital, they're trying to make the humans more productive, they're trying to make, well, make machines take over human work in some cases, applying the right labor type on to the right piece of work. Let me share a really simple example here.

CUOMO: Yes, sure, go ahead.

LYNCH: Procure to pay. Procure-to-pay is a process that most businesses have—all businesses have—because they all acquire goods. Think about Amazon's procure-to-pay. Most of us use Amazon every day. The going and looking for something in the catalogue, that's the *procure* part of procure-to-pay. Then click, buy now—that's the *automation* of Amazon's procure-to-pay.

There are no human beings in that loop at all. They are completely straight-through processing, and that's what everybody wants to do. Everybody wants to get their procure-to-pay, which, right now, employs dozens of people in every business, processing invoices, processing ERP systems, cutting checks—everybody wants to get all of those things through straight-through processing.

Why? Because it doesn't add value to the business. The procurement and payment of invoices doesn't add any value to a retailer. The retailer makes their money selling goods, and they don't want to be in the procure-to-pay business. So, really there's a consequential thing that happens when you automate people, human tasks and you let the machines do it.

That's where we are as of today.

AUTOMATION AS THE EQUAL-OPPORTUNITY EMPLOYER

CUOMO: Ed, it seems that enterprise automation can be quite the equal opportunity employer across all industries, across all job types. Is that true? Does it apply better to one than the other? What's your view on that?

LYNCH: My experience, Jerry, is that it's everywhere. It's everywhere because most of the things that are being automated are the horizontals, not the verticals. By vertical, what I mean is the sector-specific things. There's certainly a lot of work in sector-specific things like insurance claims and onboarding government documents, onboarding into different government programs, as an example. It's everywhere.

And the reason is because most of the things that are being automated are horizontals. Things like HR work, things like finance and accounting, things like onboarding mechanisms. Those things are being automated everywhere. Verticals are also being automated, but they've been automated for a long time.

People are wonder why is robotic process automation (RPA) now going completely crazy? Well, it's because we're automating things that are horizontals—like horizontal workflows that are the same no matter which sector you're in. Automation is for everyone.

LOOKING AHEAD

CUOMO: At the top of this podcast I promised the audience that there are a few people that I know who are more qualified to talk about where automation is going. So, let's talk about that now; let's talk about the future of automation. Where is it going and perhaps touch a bit on the role of AI? What role does AI play in that future?

LYNCH: I think those two thoughts are intimately tied together. We have automated. We've found mechanisms to automate lots of different types of work. But the thing that we're getting better and better and better at is augmenting human beings, making them more productive. And that's where AI and machine learning and deep learning and GANs and all the rest of these different techniques are coming into play.

They are mechanisms to transform data like raw data into patterns and then apply analytics to the patterns to give you predictions, and generate probability distributions on what's going to happen and allow individuals to make decisions based on probabilities rather than just based on one &

zero facts. This kind of Bayesian decision-making is very, very different, but it's also extremely helpful in augmenting human beings.

When a human being used to try and evaluate churn, as an example—the probability that some particular customer is going to drop your product— it used to be that you had a lot of instinct, you had a lot of observation, you had a lot of investigation. What's the past history?

Now we have churn models. A churn model is nothing more than just absorbing all the data, applying some analytical techniques, and generating a probability distribution which says the likelihood of Fred or Jerry leaving your bank is .97.

And what that does is that automatically, because of the nature of it, it allows you to automatically start doing things, the preemptive stop churn. It also applies in risk and in lending and onboarding and sentiment analysis and on and on and on. Like all of the various different things that have dozens and dozens like thousands, millions of data points, the machine can absorb them because the machines are capable of absorbing them. So, you ask what is the future of automation? I think the future of automation is AI.

CUOMO: Wonderful. I couldn't agree with you more there, Ed. Can you paint a picture of the future state of enterprise automation, maybe just continue with your thought trend? And how is it already impacting business? And can you share some key metrics to measure success by?

LYNCH: Sure. The future of automation—when you dig underneath the covers—the classification that I put on things is I say that some work is done by human labor, some work is done by digital labor. And if you think about it that way, then combining human labor and digital labor into a hybrid workforce, all of sudden a business manager can make a very active decision about where to put the work. Do I put this type of work, this task, do I give it to machine, do I give it to a human?

Now what that comes down to, that decision point comes down to what the heck are people really good at? Well, they're good at things like judgment, they're good at things like strategy, they're good at thinking out of the box. Machines are terrible at that stuff. They're terrible at judgment, they're terrible at reading emotions, they're terrible at interacting with human beings.

So, if you can make a decision as a business manager, this task is much better done by a human being than done by a machine—fantastic. Then let's figure out a way to augment the human being. If this particular task is much better done by a machine than by a human being for whatever the good reasons are, then fine, let's give it to a machine. That active hybrid workforce management is where we're headed.

And I see it emerging right now. I see it in robotics; I see it in the application of AI and machine-learning algorithms to augment human beings. So, the future state is you get to a very, very efficient back office.

And the single KPI that I keep in my mind when I think about this—I put myself in the shoes of a chief operating officer and I say, "What is that person really concerned with?" They're concerned with dollars per employee—how much revenue does the company make, divided by how many employees you have—and they're trying to continuously optimize that KPI.

CUOMO: Yes.

LYNCH: They can optimize the KPI by driving the numerator (dollars), driving loyalty, driving customer journey, driving sentiment, interaction, driving NPS, as an example. That's how you drive the numerator. You drive the denominator (employees) by making effective use of the employees that you have. You can drive the denominator to zero and get a really good dollars-per-employee, but you don't have any people and that's not going to be an effective business. So, there's a balance point.

And that future state, if you think about dollars per employee, that gives you the hint about where we're headed. They're trying to manage dollars in the numerator; sometimes dollars in the numerator requires human beings because you need people to hold hands and do customer support and that kind of thing.

CUOMO: Yes, makes sense. Ed, could you quickly weigh in on this thought? The market seems to be split and quite galvanized on business automation and IT automation. Will it continue to be that way?

LYNCH: I don't think so. I think the right north star to be thinking about is whether you're in IT or you're in business, you have people—you have human beings—and the people are doing something. If you think about the optimal way of deploying the people, whether it's in IT or in business, that's where automation comes in.

And so, whether you are optimizing the work of an operator or you're optimizing the work of a system programmer or of an SRE or of an invoice processor or somebody who's doing claims, it all comes down to the human beings.

And that human being thing—obviously human beings are extremely valuable. We don't want human beings to be completely displaced, because then you have no business. But the proper application of human beings, whether it's in IT or in business, I think it's kind of the same. And the same automation technologies come into play. And they're slightly different, obviously. You've got, in ticket management it's slightly different than processing claims. But it's still kind of the same. And so, that's where I see things going.

IT ALL STARTS WITH DATA

CUOMO: Ed, last question for you is, where do you recommend a user start to best position themselves for the future that you've just shared with us?

LYNCH: Well, I think from a user point of view, like when I talk to my clients and I say, how do you get going with this stuff? You start with the data. You have to start with the data. And the data—how rich the data is… how plentiful it is—that's going to tell you what pot of gold you're starting with.

For worker bees who are trying to get into this space, I would recommend data science. I would recommend focusing on skills that human beings are good at and making sure that those skills are on your resume. Like you've got stripes on your arms saying, yes, I'm good at this stuff.

CUOMO: Okay, Ed, that's certainly impressive and inspiring. I want to thank you for being on *The Art of Automation* today.

LYNCH: Thanks, Jerry, great talking to you.

UP NEXT

Britannica.com has an entry on automation.[42] It documents that the term "automation" was coined in the automobile industry in around 1946 to describe the increased use of automatic devices and controls in mechanized production lines. The origin of the word is attributed to D. S. Harder, an engineering manager at the Ford Motor Company at the time. It goes on to further define automation as the "application of machines to tasks once performed by human beings or, increasingly, to tasks that would otherwise be impossible."

The article warns of potential risks that automation technology will ultimately subjugate rather than serve humankind. The risks include the possibility that workers will become subservient to automated machines, that the privacy of humans will be invaded by vast computer data networks, that human error in the management of technology will somehow endanger civilization, and that society will become dependent on automation for its economic well-being.

It concludes by celebrating that these dangers aside, automation technology, if used wisely and effectively, can yield substantial opportunities for the future. There is an opportunity to relieve humans from repetitive, hazardous, and unpleasant labor in all forms. And there is an opportunity for future automation technologies to provide a growing social and economic environment in which humans can enjoy a higher standard of living and a better way of life.

Jerry and the Gang of 7 certainly agree with this conclusion. By mastering the Art of Automation to create a hybrid workforce, we hold the potential to elevate humans to a better way of life.

* * *

Make sure you check out *The Art of Automation* podcast, especially Episode 22, from which this chapter came.

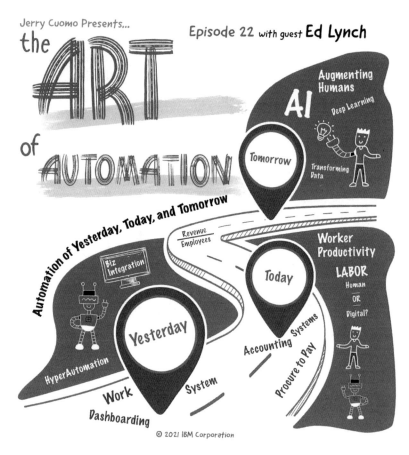

Chapter 15 – Figure 1 – Podcast Episode 22 – Automation yesterday, today, and tomorrow

Chapter 16

THE ART BEHIND
THE ART OF AUTOMATION

Pay no attention to that man behind the curtain!

Chapter Author: Ethan Glasman

COVERED IN THIS CHAPTER

- Interviews with the artists featured in *The Art of Automation* e-book

- Insights into process and artistic inspiration

- Deep dives into the meaning of each cover art

- Automation in art

BEHIND THE SCENES

Like any good television show, musical, or other pieces of media, the success and popularity of *The Art of Automation* podcast has quite a bit to do with the support staff who make the show happen from a distance. The cover art for *The Art of Automation* is recognizable anywhere. Each cover's simple—yet detailed and meaningful—art is seen by every listener and has helped establish *The Art of Automation* with its own unique personality in the world of technology podcasts.

This is by no means an accident. *The Art of Automation* has worked with some of IBM's most talented designers and visual artists to create the cover art that is now recognized by listeners across the world. Let's take this chapter to "pay attention to the people behind the curtain."

Matt Cardinal

Chapter 16 – Figure 1 – From left to right: Podcast Cover, Episode 0, Episode 1

The Art of Automation's first artist was Matt Cardinal. During the formation of the show, Matt helped design the overall podcast cover and the art for Episodes 1 and 2. Matt grew up in a number of locations, including Alberta, Arizona, and Oregon. He now lives and works for IBM in Austin. Since he can remember, Matt has constantly been interested in two things: drawing and bicycles. While these might seem like two entirely separate topics, Matt will probably disagree. In Matt's opinion, the more tactile the art, the more he enjoys it. This includes painting and, of course, welding and making bikes. In fact, he even founded his own bicycle company.

When Matt was designing his cover art, he knew he wanted something simple but impactful, especially in his choice of colors. During video calls, Matt sometimes finds himself doodling meeting participants on a sticky note, so the images of Jerry and Rama in his art came naturally. Imagery associated with the forward-thinking and futuristic nature of the podcast gave rise to the "bleep bloop!" robot as a metaphor for the AI bots used in practice. The rest of the cover started with a conversation and developed from there.

Matt's art beyond the cover art can begin from anywhere. He says inspiration isn't the difficult part—it's turning that inspiration into a final product. When making commissioned art, he actually welcomes direction and constraints, noting that it provides a place to start, while still leaving room for creativity. His process always involves a lot of trying, failing, and correcting. Once the art is complete, one of Matt's favorite things is the beauty of rediscovery. This can happen when one looks at art they haven't seen for years and finds something completely new about it, leading to newfound appreciation.

When asked if he thinks automation can ever play a role in the creation of art, he responded with "I have no idea. What does it mean to be more than an artist?" He went on to suggest that he'd like to see the creation of something that helps artists reach an end product or help remove fear from the vastness of a canvas by helping form a starting point.

Orizema Cruz Pina

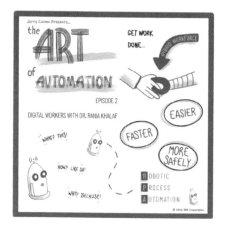 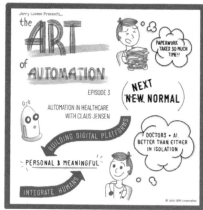

Chapter 16 – Figure 2 – From left to right: Episode 2, Episode 3

The Art of Automation's second artist was Orizema Cruz Pina. Orizema, a visual designer from San Jose, designed the covers for Episodes 2 and 3. While she's always been a creative person, doodling and decorating since she was young, she didn't consider herself an "artist" or a "designer" for many years. She discovered design in the ninth grade and was immediately drawn into the structure of it. She loves doing digital art but lately has found a bit of a balance by expanding her interests to include painting, charcoal, and watercolor.

When Orizema took on the task of creating cover art for the podcast, she thought of it as a fun challenge to design within a system, where color and style had already been established by Matt. She loved building off what he made in the first few episodes while adding in her own flair. She had never done art in this style, but she knew she wanted to make it friendly. Just like the podcast itself, the idea of Orizema's art was intended to be consumable, even for someone who was just beginning their journey in the AI and automation world. Images like the "robot handshake" and the "stack of paper guy" are designed to be fun while still conveying the main points of that podcast episode.

Orizema finds artistic inspiration from her own experiences in life. This includes her cat or things she's seen in nature. She doesn't really enjoy trying to realistically draw people, but this has manifested in a new passion for cartoons and comics, as is evident in the podcast art she made. Her process always begins with a pencil and paper, loosely drawing out sketches and ideas. This is followed by tracing and layering, which leads to moving around different images to get the ideal layout. This happened quite a bit in the Episode 3 cover art, where the two characters and the robot moved around at least three times each. Her final step is making the text more fun and cleaning everything up.

When asked if she thinks automation can ever play a role in the creation of art, Orizema responded by saying that automation could best be used to improve what exists today. This could take the form of better features in a graphic design program, such as perfecting shapes and lines. Overall, she sees automation as a tool that designers and artists can use to help make their visions come to life.

April Monson

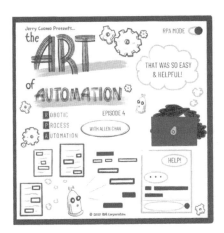 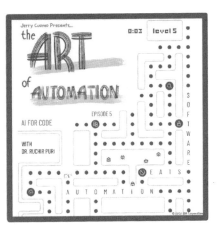

Chapter 16 – Figure 3 – From left to right: Episode 4, Episode 5

The Art of Automation's third artist was April Monson. April designed the covers for Episodes 4 and 5. She is a user experience designer from San Diego who attributes her interest in visual design to her childhood as an "indoor kid." Her brother was also interested in visual design, so it turned into a form of competition, which still slightly exists today. April's mother was also very supportive of her and her brother's passion and fostered it by taking them to museums and encouraging their artistic drives.

When designing the cover art for *The Art of Automation*, April said she was a little nervous at first, as she did not consider herself an expert visual designer. So, her first piece of art on robotic process automation (RPA) was essentially done in the style of how she takes notes during lectures or all-hands meetings. To April, drawing small images and diagrams makes things stand out and stick to one's memory, which is what she tried to capture in the art for this episode. She wanted to convey the idea that software robots are your friends, there to help make your work easier. For the next cover art, however, April wanted to do something a little more unique. She was inspired by the quote "automation eats software," as well as the imagery of Russian dolls and Pac-Man. She likes the idea of taking up more whitespace than previous art and showing how different levels of automation can help you find context/improve/learn to become more helpful.

In general, April likes to find inspiration for her art by walking in nature, sitting on her balcony, or listening to music. As a first step in creation, she likes to journal or doodle on a topic to see what comes up. A lot of her art is reflective of what she's feeling or what she did that day. She loves art that is "in the moment." As such, some of her favorite artwork is done simply with a pencil and paper because it is cheap, easy, and she has all sort of journals lying around. She also very much enjoys knitting and pyrography, which is the art of decorating wood with burn marks from a heated metallic burning pen.

When asked about the relationship between automation and art, April had several interesting ideas for what she'd like to see in the future. One technology that she's seen early versions of involves entering a phrase or an idea as text and having AI generate an image (that fits your project style) to be used in your art. She was also interested in the idea of using AI and automation to learn about famous paintings or art and translating those insights into tools or suggestions that could be used by artists today.

Caroline Scholer

Chapter 16 – Figure 4 – Episode 7

The Art of Automation's fourth artist was Caroline Scholer. Caroline designed the cover for Episode 7 on Intelligent Document Automation. She is a student at University of North Carolina at Asheville, North Carolina, who took a serious interest in art at eleven years old through reading and cartoons. Her favorite kind of art is painting, but she also enjoys working on digital platforms from time to time. Caroline enjoys painting the most because it allows her to experiment with mediums and gives her opportunities to grow artistically. She particularly enjoys the process of painting and seeing a project progress step by step.

When designing for *The Art of Automation*, Caroline's process began with a few concept sketches. This was a way for her to pilot several ideas for general design and decide what specific images she wanted to include. Her next step was taking a photo of her final sketch and tracing over it digitally. She then altered the color scheme and adjusted placement to get a layout she liked. Her final step was to set everything and add *The Art of Automation* watermarks.

Outside of creating cover art, Caroline is inspired for art by the idea of incorporating emotions and experiences into something visual, allowing others to experience it as well. She also likes to be inspired by her fellow artists, taking observations and insights from other creators, and incorporating them into her own work. In general, Caroline believes that art is based on someone's interpretation of an idea, and she finds it fascinating to be able to share her perspectives with others.

When asked about how art and automation can work together, Caroline remarked that "automation allows for a faster and more structured execution of an artist's ideas." She elaborated that when barriers are eliminated in the artistic process, it allows for more genuine and efficient work. This can be as simple as mechanical pencils saving time on sharpening, to advanced AI tools helping speed up other processes. She firmly believes that automation has the ability to help artists and will continue to do so in a greater capacity as technology evolves.

Danielle Elchik

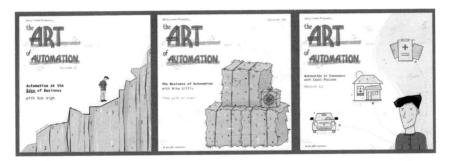

Chapter 16 - Figure 5 - From left to right: Episode 8, Episode 10, Episode 12

The Art of Automation's fifth artist was Danielle Elchik. Danielle designed the covers for Episodes 8, 10, and 12. She is a graphic and visual designer from Pittsburgh who has been interested in art and creativity from a very young age. In her free time, she likes to do physical, tangible art (such as drawing or painting), but professionally, she focuses on digital art. Although, she always enjoys when her work is printed onto actual paper, such as when she wrote and designed for the IBM Variable Magazine.

Her process for designing the cover art of *The Art of Automation* began with listening to each episode and taking a few notes. From there she would use individual quotes and try to identify an overall concept or message that could be transformed into some sort of powerful visual. The next step for Danielle was making thumbnails and rough sketches and messing around with the layout.

One goal she had when designing the art was to not have everything make complete sense from one's first look. For example, when first looking at the Episode 8 artwork of a person standing at the edge of the cliff, the meaning is not immediately clear. However, after listening to the podcast and learning more, one can easily come back to the visual and find a clearer meaning relating back to the main idea of pushing automation to edge devices in a business (and other topics covered in the episode).

It's this sort of abstract interpretation that Danielle enjoys, which can similarly be seen in her Episode 10 artwork, combining "finding a needle in a haystack" and "the gift of time." She notes that she especially enjoyed playing with texture and shadowing in that cover art. Additionally, all three of her pieces of art contain datapoints and numbers to fill out the white space and illustrate the idea that data in coming from everywhere. Automation, Danielle says, always comes back to the data that fuels it.

Artistic inspiration for Danielle comes from all sorts of places, but she particularly likes browsing online designer collaboration sites. This allows her to learn from other artists' styles and investigate how they are telling stories and communicating through their artwork. Danielle is very interested in this practice of telling stories through design and visual art. To see how talented she is at doing so, one needs to look no further than the IBM Variable Magazine mentioned above, where Danielle was instrumental in the creation of pieces such as "The Unconscious Mind" and "Welcome to the Savage."

Danielle's first thought about incorporating automation into the world of visual art was that it should be used as it is in other fields—to help humas focus on more meaningful tasks by automating the repetitive and boring ones. One idea she had was to help artists with the experimentation process, to help them discover what they're looking for. Perhaps an AI model could give you fifty different variations of one of your drafts, with a variety of layout and pixilation options to inspire you for your next steps. Or, when creating something like a slide deck, design and layout suggestions can be massively beneficial to someone who is not as confident in their artistic skills. She also suggested a number of ways automation could help in the user experience design process, by automating tedious tasks to allow you to focus on addressing customer pain points and needs. In this scenario, humans are still doing the problem solving, but automation is helping speed up the execution.

Adaoha Onyekwelu

 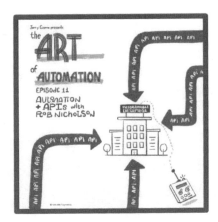

Chapter 16 – Figure 6 – From left to right: Episode 9, Episode 11

The Art of Automation's sixth artist was Adaoha Onyekwelu. Adaoha designed the covers for Episodes 9 and 11. She is a UX designer from New Jersey who spent five years of her childhood living in Nigeria. She became interested in visual design during elementary school and always remembers having notebooks dedicated to drawing. While she enjoys drawing on paper, most of her work is done digitally on a tablet. Although, some of her favorite projects were done entirely by hand.

Her process for creating cover art for *The Art of Automation* began with listening to the episode a few times and trying to pick out key words. After getting all the words on paper, Adaoha's next step was word association, where she brainstormed any images, items, or objects that related in her mind to the key words. For Episode 9, an obvious keyword was "observable," which gave rise to images of eyes and a magnifying glass. After she decided on the images she wanted to use, she solidified the visuals by sketching them out and organizing them in a desired layout. Finally, she digitized everything and added finishing touches.

While Adaoha enjoys cover art design and UX design, her favorite form of visual art is hand lettering. Hand lettering, as the name suggests, is an art form that centers around letters and words to convey images and ideas. Adaoha discovered hand lettering a few years ago and has since been intrigued by it nonstop. As she puts it, "Hand lettering is so freeing," mentioning that she likes hand lettering so much because it is not restrained by the usual formal art rules of colors and structures. In hand lettering, anything goes; the only limitation is your imagination. She now enjoys digital lettering the most because she feels like one can do so much more on a digital platform. Although once in a while, she will go back to brush, pen, and paper. Some of her recent personal projects include wedding signs and hand menus, and she is having a lot of fun establishing her own artistic style.

When asked how automation can play a role in art, Adaoha's first thoughts were around finding artistic inspiration. She is interested in how technology can play a role in inspiring artists, whether that might be learning their style and making suggestions for a new project or providing inspiration for where to take a project next. She then suggested that technology can go even further, perhaps helping artists in their creatives processes in ways we can't even imagine yet.

David Ryan

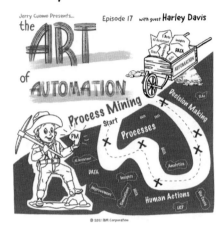

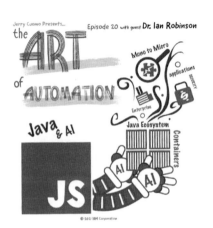

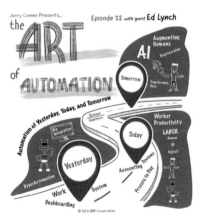

Chapter 16 – Figure 8 – David Ryan's cover art portfolio

The Art of Automation's seventh artist was David Ryan. David designed the covers of Episodes 17, 18, 20, 21, and 22. He is an artist originally from Michigan and currently based out of North Carolina. While he is widely considered a brilliant visual artist, his favorite form of art is actually not visual at all—it's music. A video and graphic designer by day, David plays guitar in a rock band during his free time. He also loves sci-fi animation, especially when it's designed specifically to go along with a piece of music.

When designing for *The Art of Automation*, David recalls, "I listen to the podcast, and within the first ten seconds of the recording, I usually know what the theme of the visual will be." These themes have been a hit with *The Art of Automation's* audience. Images like "The Prospector" from Episode 17 and "The Java Robot" from Episode 20 give each episode character and flavor that stick in the minds of listeners. David comes from a family of artists and considers his childhood and upbringing critical to the development of the artist he is today.

When asked if he believes that automation can ever play a role in the creation of visual art, David explained that he's always believed AI and automation can be "smart enough" to create design solutions. Of course, they'll need plenty of visual and creative problems to train on before that's possible. He elaborates that automation may be powerful enough to do this in theory, but he has yet to see it in practice.

There is no better way to close this chapter than with a quote from David himself: "Art should comfort the disturbed and disturb the comfortable. I hope my visual art does that. If not, it's not successful."

Thank you to all of our featured artists. It would not be the "Art" of Automation without you.

Chapter 17

Ciao

Wait... you're still here? The book is done.

Well, since you're still reading, let's share a few closing thoughts.

Chapter Author: Jerry Cuomo

COVERED IN THE CHAPTER

- Wrap-up

- Not in this book

- "DJ" on LinkedIn

- *The Art of Automation* Season 2 preview

WRAP-UP

The word "ciao" is mostly used as "goodbye" or "bye" in English, but in modern Italian and in other languages it may also mean "hello." Ciao captures the spirit of this short chapter. It is a bit of a "bye for now" while also providing a "hello" and a means to stay in touch. Playing on that sentiment, this chapter covers a few loose ends, including topics that were not covered in this book, a pointer and an overview of *The Art of Automation*

LinkedIn page, and a preview of *The Art of Automation* podcast Season 2. Thanks again for your continued support of The Art of Automation project: a project that explores the "art of the possible" of applying AI-powered automation in the enterprise, which can now be experienced as a podcast, a book, and on social media.

NOT IN THIS BOOK

There are only so many hours in a day. There are a few topics that we wish we had time to cover but unfortunately did not. Some of these topics have already been featured as episodes of *The Art of Automation* podcast. What follows is a quick enumeration of these topics, with a short description and pointer to more information on them. Perhaps these topics will be covered in the second edition of this book.

AI for Code

For Episode 5 of the podcast, Jerry is joined by IBM Fellow, chief scientist of IBM Research, and one of the architects behind Watson (of *Jeopardy!* fame), Dr. Ruchir Puri. They discuss why it's important to "teach machines their own language" and what automating the creation of software and IT processes can mean for an enterprise with millions of lines of code. Ruchir elaborates on the daily benefits this automation has for developers, from code search to code translation to modernizing legacy code.[43] [44]

Automation with Java

For Episode 20 of the podcast, Jerry is joined by IBM distinguished engineer, CTO of IBM Application Platforms, and a founding father of the WebSphere Application Platform, Dr. Ian Robinson. Jerry and Ian discuss the story of Java, from its beginnings at Sun Microsystems to its role today as the platform for an enduring software development ecosystem of languages, runtime platforms, open-source projects, and much more. Ian explains how the internet as we know it today would simply never have existed without Java. He then elaborates on automation and AI's role in Java with examples from real operations, IT, and other enterprise teams. They close by exploring a new game-changing technology (which combines Java and AI-powered automation) that is taking monolithic applications and increasing their agility by breaking them down into smaller microservices.[45 46 47 48]

Automation and Hybrid Cloud

For Episode 15 of the podcast, Jerry is joined by IBM distinguished engineer, CTO of IBM Cloud Paks, and the host of The Hybrid Cloud Forecast podcast, Andre Tost. They begin by discussing a broad definition of hybrid cloud, which includes flexibility of location and the ability to support a wide variety of application architectures. Andre shares his view that automation is one of the core principles of cloud and that it's impossible to do cloud computing without it. He also discusses how AI brings intelligence to automating hybrid cloud, by helping predict and fix operational problems before they happen.

Automation at the Edge

For Episode 8 of the podcast, Jerry is joined by IBM fellow and VP/CTO of IBM Edge Computing, Rob High. They discuss how edge computing keeps businesses relevant in the "digital NOW age," by lowering latency in decision-making and data processing, reducing bandwidth, and protecting sensitive information. Rob describes what it means to bring automation out to where the data is collected ("the edge"), such as machines on a factory floor, and why doing so can be massively beneficial for an enterprise.

Automation in Blockchain

For Episode 19 of the podcast, Digital Jerry (DJ) and Human Jerry return for a special episode on one of the most popular technology trends of the last decade, Blockchain. Jerry begins by clearing up a few common misconceptions about Blockchain, like the fact that it's not the same thing as Bitcoin. He reframes the popular narrative around Blockchain by sharing its extensive use cases, all of which are unified by a desire to build a trusted network to handle interactions between known parties. Jerry then shifts the discussion to enterprise AI-powered automation, where automation technology is generally bounded by a company's firewall. With Blockchain, however, automation is being expanded to multi-party processes (such as supply chain), saving substantial amounts of time and money. Jerry closes by summarizing how Blockchain's immutable shared ledger creates a foundation of trust and confidence, by forming one system of truth for all players involved.[49]

Automation Architecture

For Episode 13, Jerry is joined by IBM distinguished engineer and lead architect of the IBM Automation Foundation, Pratik Gupta. They discuss what automation is made of and how the blueprint of automation can be broken down into "Discover, Decide, Act, and Optimize." Pratik shares an example of how automation has evolved in manufacturing cars, from the Ford Model T to the Tesla Model Y. He ties this back to enterprise automation with a use case from the information technology space, where decisions are constantly being made, for hundreds of thousands of people, automatically or—as Pratik puts it—"Planet-Scale AI-Powered Automation."

Automation in Additional Industries

This will be covered in the "Season 2 preview" section that follows.

WHO IS D.J.?

D.J. is an abbreviation for Digital Jerry. D.J. has made several appearances on *The Art of Automation* podcast, including Episode 6 – 2021 Predictions as well as several bonus episodes. The D.J. persona is powered by a high-quality text-to-speech engine, enabling it to perform interviews with Human Jerry during podcasts. D.J. plays well into the automation theme and hopes someday to evolve into a digital employee.

ART OF AUTOMATION ON LINKEDIN

We've created a LinkedIn account, Art of Automation – DJ, as a landing spot for content related to the book as well as to provide a future means to interact and get feedback from podcast listeners and book readers. The account features articles and references to automation topics covered both in this book and podcast episodes. This account will also allow us to solicit feedback in the form of surveys and comments. To connect to our LinkedIn account, simply follow the link below, and as the following figure illustrates, click the "Connect" button to join our network.

https://www.linkedin.com/in/art-of-automation/

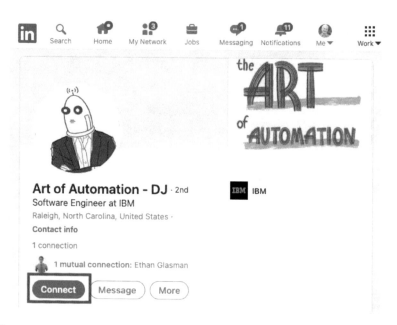

Chapter 17 – Figure 1 – Connecting to Art of Automation – DJ page on LinkedIn

THE ART OF AUTOMATION SEASON 2 PREVIEW

Season 1 of *The Art of Automation* podcast spanned the course of a year (October 2020–October 2021). The season comprised twenty-three episodes that were downloaded over eleven thousand times across ninety-one countries… and counting. The theme for Season 1 was a mix of technology and industry topics; however, it was more skewed towards technology. Season 2 hopes to flip that around and focus more on industry applications of AI-powered automation.

Specific industries that we hope to cover include *manufacturing, food, automotive,* and *media and entertainment.* The technology topics listed in the previous section, plus coverage of these new industry areas, could fuel the foundation for a second edition of this book or a Volume 2.

But that's not something that I would like to consider on my own. I want to stay in touch with you and hear your ideas. So please share your feedback by connecting to our LinkedIn account as described.

* * *

Well... that's all folks!

ACKNOWLEDGEMENTS

Foreword Author

>> *April 1, 2020:* **Mike Gilfix:** *Can you talk? I had a major epiphany. Like string theory... it all clicked.*

That was the Slack message that I saw at 2:34 p.m. ET. So, I quickly called Mike. During the call, Mike suggested we look at the automation opportunity concisely and end to end, from business, software delivery, and ITOps. And here we are.... Thanks for bringing the pieces of the puzzle together.

Michael Gilfix is the vice president for AI-powered automation products at IBM. Mike has a decorated history where he's held roles in engineering, architecture, business development, and later, as product executive of mobile offerings, and then taking the helm of our business automation portfolio in 2016.

Industry Experts

Thank you to the experts that participated in our industry-focused interviews, in Chapters 9–15. Your examples and insights clearly show how this "art" can be put to work for positive change.

Claus Jensen, chief digital officer at Memorial Sloan Kettering Cancer Center, on automation in the healthcare industry.

Carol Poulsen, chief information officer at The Co-operators Group, on automation in the Insurance Industry.

Tim Vanderham, chief technology officer at NCR Corporation, on automation in the retail industry.

Oscar Roque, VP of Strategy at Interac Corporation, on automation in the financial services industry.

Lisa Seacat DeLuca, distinguished engineer, author, and one of the most prolific inventors in the history of IBM, on automation and the weather.

Don Scott, director of engineering at Submergence Group and the mastermind behind the *Mayflower Autonomous Ship,* on automation at sea.

Ed Lynch, VP of business automation at IBM, on the past, present, and future of enterprise automation.

Reviewers

The co-authors would like to thank the following colleagues for reviewing and editing the contents of this book:

Ian Smalley, our webmaster and blog editor since the inception of this project. Ian helped us pilot the book on IBM.com, unveiling chapters as they became available.

Barry Mosakowski, our most diligent reviewer, who decided the best way to provide feedback on the book was to call Jerry each morning (during the review process) and give a personal recap of his (always insightful) suggestions for improvement.

Diane Hatcher, another of our diligent reviewers, who agreed to write the first Amazon review of this book. Good or bad!

Bill Lobig, Thomas Edison once said that vision without execution is just hallucination. To the guy that made IBM Automation a reality.

Haechul Shin, who helped formulate IBM's AI-powered automation strategy and the Automation 2.0 perspectives shared in Chapter 1.

Bobbie Cochrane, who resisted grammar checking this book and instead provided thorough feedback on the AIOps and observability chapters.

Mark Parzygnat, who often provided technical and mental support over beer and wings.

Gale Fletcher, who ensured that there was always room on Jerry's calendar for recording *The Art of Automation* podcast.

The authors also want to acknowledge the help and support from Memsy Price, Liz Urheim, Brandy Hartford and Jonathan Young.

References

Foreword

1 – Cuomo, Jerry. "The Art of Automation." Spotify. October 2020. *https://open.spotify.com/show/4CLOddf8rBxTnO3Hw1DKEX*

2 – "The Business of Value of Using IBM AI-Powered Automation Solutions." International Data Corporation. August 2021. https://www.ibm.com/downloads/cas/BBNZ1KLW

3 – Barnett, Jackson. "By using AI, the VA dramatically decreased claims processing intake times, official says." Fedscoop. July 1, 2020. https://www.fedscoop.com/veterans-benefits-ai-mail-processing/

Preface

4 – "What Is Digital Transformation?" Salesforce.com. https://www.salesforce.com/products/platform/what-is-digital-transformation/

5 – Design Patterns – The Gang of 4. Wikipedia. https://en.wikipedia.org/wiki/Design_Patterns

Chapter 1 – Introduction

6 – Jesuthasan, Ravin and Boudreau, John. "How to Break Down Work into Tasks that Can Be Automated." Harvard Business Review. February 20, 2019.

https://hbr.org/2019/02/how-to-break-down-work-into-tasks-that-can-be-automated

7 – "Gartner Top Strategic Technology Trends for 2021." Gartner Inc. October 19, 2020.

https://www.gartner.com/smarterwithgartner/gartner-top-strategic-technology-trends-for-2021

8 – Gelles, David and Corkery, Michael. "Robots welcome to take

over, as coronavirus pandemic accelerates automation." Seattle Times. April 12, 2020. https://www.seattletimes.com/ business/robots-welcome-to-take-over-as-coronavirus-pandemic-accelerates-automation/

9 – Maurer M., Breskovic I., Emeakaroha V. C., et al. "Revealing the MAPE loop for the autonomic management of Cloud infrastructures." IEEE Xplore (2011). 147–52. https://ieeexplore.ieee.org/document/5984008

10 – Taylor, Christine. "Structured vs. Unstructured Data." Datamation. May 21, 2021. https://www.datamation.com/big-data/structured-vs-unstructured-data/

11 – Kavlakoglu, Eda. "AI vs. Machine Learning vs. Deep Learning vs. Neural Networks: What's the Difference?" Cloud Education, IBM. May 27, 2020. https://www.ibm.com/cloud/blog/ai-vs-machine-learning-vs-deep-learning-vs-neural-networks

12 – Prasad, Pankaj and Rich, Charley. "Market Guide for AIOps Platforms." Gartner Inc. November 12, 2018.

https://www.gartner.com/en/documents/3892967/market-guide-for-aiops-platforms

13 – "The evolution of process automation." IBM Institute for Business (2018). https://www.ibm.com/thought-leadership/institute-business-value/report/ibvprocessautomation

14 – American Productivity and Quality Center (APQC). APQC. https://www.apqc.org/resource-library

15 – Allen, Katie. "Technology has created more jobs than it has destroyed, says 140 years of data." The Guardian. August 18, 2015. https://www.theguardian.com/business/2015/aug/17/technology-created-more-jobs-than-destroyed-140-years-data-census

Chapter 2 – Robotic Process Automation (RPA)

16 – Brain, David. "RPA Technical Insights, Part 10: Why Screen Scraping is Essential to the RPA Toolkit." December 8, 2016. https://www.symphonyhq.com/rpa-technical-insights-part-10/

17 – Attended and Unattended RPA, Explained. Automation Anywhere. https://www.automationanywhere.com/rpa/attended-vs-unattended-rpa

18 – IBM. "Demo Overview of IBM Robotic Process Automation (RPA)." YouTube video, 4:09. March 12, 2021. https://www.youtube.com/watch?v=CRaI8zhops0

Chapter 3 – Process Mining

19 – Process Mining. IBM Cloud Education. January 8, 2021. https://www.ibm.com/cloud/learn/process-mining

Chapter 4 – Digital Employees

20 – Free, Andrew. Conversational AI: Chatbots that Work. Manning Publications, 2021. https://www.manning.com/books/conversational-ai

21 – "Intelligent Document Processing Market by Component (Solutions, Services), Deployment Mode (Cloud, On-Premises), Organization Size, Technology, Vertical (BFSI, Government, Healthcare and Life Sciences), and Region – Global Forecast to 2026." Markets and Markets. https://www.marketsandmarkets.com/Market-Reports/intelligent-document-processing-market-195513136.html

Chapter 5 – Intelligent Document Processing

22 – Marr, Bernard. "How Much Data Do We Create Every Day? The Mind-Blowing Stats Everyone Should Read." Forbes. May 21, 2018. https://www.forbes.com/sites/bernardmarr/2018/05/21/how-much-data-do-we-create-every-day-the-mind-blowing-stats-everyone-should-read/

23 – "2019 Payables Insight Report: Understanding the Value

of Holistic Invoice-to-Payment Automation for Enhanced Business Outcomes." Levvel Research. https://www. expenseanywhere.com/wp-content/uploads/2019/04/2019-Payables-Insight-Report_ExpenseAnywhere.pdf

24 – Jenness, David. "IBM Automation Document Processing – Demonstration of a Utility Bill Paying Application." YouTube video. 4:00. March 25, 2021. https://www.youtube.com/watch?v=Df05HoJ3Lr8

Chapter 6 – Observability

25 – What Is Observability? IBM Cloud Education. January 12, 2021. https://www.ibm.com/cloud/learn/observability

26 – Murphy, Niall. "Site Reliability Engineering – Ben Treynor." Google. https://sre.google/in-conversation/

27 – Instana, IBM. "Instana Quick Overview." YouTube video, 4:07. May 8, 2021. https://www.youtube.com/watch?v=7Q4Lllk6OJM

Chapter 7 – AIOps

28 – IT Operations. IBM Cloud Education. December 10, 2020. https://www.ibm.com/cloud/learn/it-operations

29 – Xiaotong, L. et al. "Using Language Models to Pre-train Features for Optimizing Information Technology Operations Management Tasks." International Workshop on Artificial Intelligence for IT Operations (AIOps) (2020). https://aiopsworkshop.github.io/accepted_papers/index.html

30 – Devlin, J. et al. "BERT: Pre-training of Deep Bidirectional Transformers for Language Understanding." Association for Computational Linguistics (2018). 4171–186. https://arxiv.org/abs/1810.04805

31 – Rezaeinia, S. M. et al. "Sentiment Analysis Based on Improved Pre-trained Word Embeddings." Expert Systems with Applications (2018). 139–47. https://doi.org/10.1016/j.

eswa.2018.08.044

32 – Liu, Xiaotong, Xu, Anbang and Akkiraju, Rama. "Using Language Models to Optimize IT Operations Management." IBM. December 8, 2020. https://www.ibm.com/cloud/blog/using-language-models-to-optimize-it-operations-management-in-watson-aiops

33 – Akbik, A. et al. "Generating High Quality Proposition Banks for Multilingual Semantic Role Labeling." Association for Computational Linguistics (2015). 397–407. https://aclanthology.org/P15-1039.pdf

34 – Aggarwal, P. et al. "Mining Domain-Specific Component-Action Links for Technical Support Documents." CODS-COMAD (2021). 323–31. https://dl.acm.org/doi/abs/10.1145/3430984.3431000

35 – "The Bezos API Mandate: Amazon's Manifesto for Externalization." Nordic APIs. January 19, 2021. https://nordicapis.com/the-bezos-api-mandate-amazons-manifesto-for-externalization/

Chapter 8 – APIs

36 –Kosinski, Matthew. "What Is Closed-Loop Integration?" IBM. September 8, 2021.

https://www.ibm.com/cloud/blog/what-is-closed-loop-integration/

Chapter 9 – Healthcare

37 – Williams, Paula. "Amazing Ways That RPA Can Be Used in Healthcare." IBM. September 21, 2021. https://www.ibm.com/cloud/blog/amazing-ways-that-rpa-can-be-used-in-healthcare

Chapter 11 – Retail

38 – Begley, Steven et al. "Automation in Retail: An Executive Overview for Getting Ready." McKinsey & Company. May

23, 2019. https://www.mckinsey.com/industries/retail/our-insights/automation-in-retail-an-executive-overview-for-getting-ready

Chapter 13 – Automation and the Weather

39 – DeLuca, L. S., and Greenberger, J. A. "Cognitive geofence updates." IBM. September 22, 2020. https://patents.google.com/patent/US10785598B2/

Chapter 14 – Automation at Sea

40 – Mayflower Autonomous Ship. https://mas400.com/

41 – Smith, Robert. "The Key Differences between Rule-Based AI and Machine Learning." Becoming Human –Artificial Intelligence. July 14, 2020. https://becominghuman.ai/the-key-differences-between-rule-based-ai-and-machine-learning-8792e545e6

Chapter 15 – Yesterday, Today, and Tomorrow

42 – Groover, M. P. "Automation." Britannica.com. October 22, 2020. https://www.britannica.com/technology/automation

Chapter 17 – Ciao

43 – More on ML for Code. https://ml4code.github.io

44 – More on modernizing legacy code. https://spectrum.ieee.org/tech-talk/artificial-intelligence/machine-learning/ibm-ai-watson-modernize-legacy-code.

45 – IBM Semeru Runtimes. http://ibm.com/semeru-runtimes

46 – The Open Liberty Project for Java microservices. An IBM Opensource project. https://openliberty.io/

47 – Mono2Micro. IBM. http://ibm.biz/Mono2Micro

48 –WebSphere Automation. IBM WebSphere. https://www.ibm.com/cloud/websphere-automation

49 – Cuomo, Jerry. "In Data We Trust. Well… after You Add a Little Blockchain." LinkedIn Blog. October 21, 2019. https://www.linkedin.com/pulse/data-we-trust-well-after-you-add-little-blockchain-jerry-cuomo/